Advancing Physics AS

British Library Cataloguing in Publication Data
A catalogue record for this book is available from the British Library

First published June 2000
Revised edition published 2008
by Institute of Physics Publishing, Dirac House, Temple Back, Bristol BS1 6BE, UK

ISBN 978-0-7503-0585-3

Printed and bound in Spain by Mateu Cromo, Pinto (Madrid)

Revised and first edition authors and contributors (books and CD-ROMs):

Steve Adams,
Susan Aldridge
Michael Barnett
Christopher Bishop
Richard Boohan
David Brabban
Jim Breithaupt (A–Z)
Michael Brimicombe
Philip Britton
Ian Brown
James Butler
Peter Campbell
Geoff Camplin
Simon Carson
Simon Collins
Mark Cramoysan
John Cullerne
Kathleen Davies
Laurence Dickie
Ken Dobson

Ingrid Ebeyer
Tony Egan
Diana Emes
Jonathan Foyle
Richard Field
Mike Gidlow
David Grace
Sarah Grant
Stephen Hall
Stephen Hearn
Lawrence Herklots
David Homer
Julian Hoult
Neil Hutton
Mike Kearney
Ian Lawrence
Allan Mann
Rick Marshall
John Mascall
John Miller

John Mitchell
David Morland
Andrew Morrison
Jon Ogborn (project director)
Simon Petts
Steve Pickersgill
Andrew Raw
Helen Reynolds
Laurence Rogers
David Sang
Clare Sansom
Allan Seago
Chris Shilliday
Robert Strawson
Janet Taylor
Kevin Walsh
Elizabeth Weiser
Catherine Wilson
Ken Zetie

Revised edition production:

Susannah Bruce, Laura Churchill, Kate Gardner, Andrew Giaquinto, Jane Henley, Kerry Hopkins,
Anastasia Ireland, Huw Johnson, Evie Palmer, Cee Pike, Teresa Ryan, Andrew Stevens, Fred Swist,
Jamie Symonds, Alison Tovey

First edition production:

Penelope Barber, Alan Evans, Huw Johnson, Hayley Liddle, Kevin Lowry, Bridget Pairaudeau, Angela
Gage, Andrew Stevens, Jamie Symonds, Alison Tovey, Brenda Trigg, Lucy Williams

ADVANCING PHYSICS AS

Revised edition edited by
Jon Ogborn and Rick Marshall

First edition edited by
Jon Ogborn and Mary Whitehouse

CD-ROM edited by
Ian Lawrence, Rick Marshall and Jon Ogborn

CD-ROM first edition edited by
Ian Lawrence and Mary Whitehouse

IOP Publishing

Contents

How to use this book and CD-ROM

The book and CD-ROM go together. Both are essential to the Advancing Physics course.

The book provides, in each chapter:
- **Narrative text** telling the story of part of physics, putting it in context and explaining to you why it is worth understanding
- **Key summaries** that contain all the essential ideas, showing their structure in a visual form
- Short **Quick check** questions at the end of each section, to start you off being able to make calculations and arguments for yourself
- **Links to the CD-ROM** at the end of each section, pointing out further questions for practice, activities to try, key terms to look up in the A–Z, further readings for interest and key items to use for revision
- At the end of the chapter a **Summary check-up** lists all the essential ideas you should have learned, with some further **Questions** to try.

The CD-ROM provides, for each chapter:
- **Activities**, including some software on the CD-ROM, and many experiments to do
- **Questions** at different levels, from warm-up and practice exercises to questions at AS examination level
- **Readings** that extend the material in the textbook or provide interesting alternative views of the physics in the course and its background, with an emphasis on the people involved
- **Images** to look at that add to those in the textbook, and **Key summary** diagrams to study
- **Computer files** for use with a variety of computer software, notably spreadsheets and modelling packages
- **Revision checklists** of things you should have learned to be able to do
- **An A–Z of physics** covering all the ideas in the course in a form useful for revision
- **Help with skills** of experimenting and data handling

Studying a section of a chapter ✓

Each chapter is divided into numbered sections. When studying a section you should:
- First read the section through, get a sense of what it is all about
- Work through lab activities on the CD-ROM, guided by your teacher
- Go back and study the Key summaries, which display ideas pictorially. You will have to work at these to get their full message. Look carefully at how the logic of arguments and relationships between ideas are laid out graphically. Try reconstructing the diagrams for yourself – this can teach you a lot.
- Work through all of the Quick check questions
- Follow the Links to the CD-ROM for more questions to try, activities to look at, key terms to check out and other items for further interest or revision
- Use the A–Z to look up ideas you don't understand well enough, or to get a view different from the one in the book
- Personalise your CD-ROM by creating your own "shadow file" of notes, comments, summaries, etc.

Revising for tests or examinations ✓

- Review your "shadow file" notes
- Go to the CD-ROM revision checklist for each chapter and follow the links to relevant revision terms
- Browse the A–Z terms listed for each section of a chapter
- Review the relevant Key summary diagrams in the book or on the CD-ROM
- Try more questions on the CD-ROM
- Stretch yourself with some further readings
- Try further activities including ones using software

The Revision CD that is also available provides
- Checklists of things you need to know and be able to do
- Revision notes and diagrams
- Previous examination questions
- Worked examples

Links between the student's book and CD-ROM

At the end of each section of each chapter you will find a list of links to the CD-ROM that provide one-click access to:

- questions to practise with
- activities to try out
- key terms to look up in the A–Z
- readings to go further for interest
- the revision checklist and other revision resources

Book page

Quick check

CD links

Book summary check-up page

CD-ROM page

1 Imaging

Visualising data is an important and rapidly developing aspect of physics. In this chapter we will give examples of images for:

- use in medicine
- exploring the universe
- observing the Earth from space
- "seeing" atoms

The examples will lead you to new ideas about processing and measuring information, and about vision and optical instruments.

1.1 Seeing invisible things

Ultrasound imaging in pregnancy

The scene is the antenatal department of a hospital. A young woman is there for a check-up. An image of the baby inside her appears on a computer screen. Is everything all right? Seeing a baby develop inside its mother is no longer considered remarkable. Ultrasound scanning is routinely used as a check during pregnancy. Is scanning safe? We have no evidence that these scans harm an unborn baby, but having no evidence of an effect does not prove that there is no effect. By contrast, the potentially damaging effects of X-rays are well known. Doctors use ultrasound scans rather than X-rays where possible, but scans are still used with caution, and mothers may

Ultimate reality

A young man said to me:
I am interested in the problem of reality.

I said: Really!
Then I saw him turn to glance,
surreptitiously,
In the big mirror, at his own fascinating shadow.

D H Lawrence, The Complete Poems (1994) Penguin

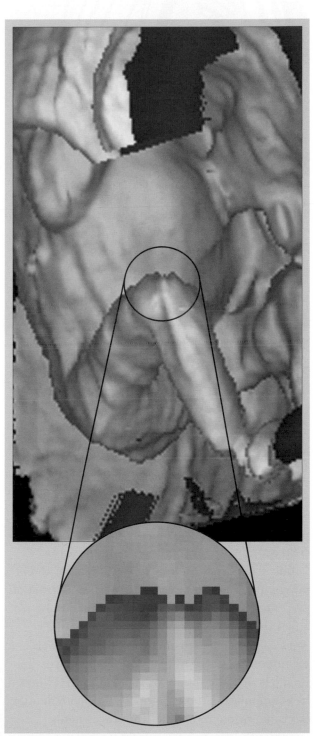

An ultrasound image of a 20-week-old foetus. The baby's legs are clearly visible. The head is obscured by part of the wall of the womb. The sound frequency used is several millions of oscillations per second — megahertz — which is much too high to be audible by the human ear. The enlargement shows a region of the baby's right knee. The discrete square pixels are clearly visible, as are the distinct levels of grey.

If you look at this painting by Lichtenstein under a magnifying glass you can see red dots of paint. At a distance, your eyes cannot resolve the dots so you see a continuous colour.

27	28	33	39	168	239	47
27	242	34	235	150	153	218
121	138	213	179	151	146	171
104	83	146	148	124	102	116
32	57	94	113	89	90	113
31	46	62	105	87	86	91
31	44	50	59	76	77	75

The number for each pixel decides what shade of grey (or what colour) that pixel will be printed. You can see how darker regions in the image correspond to higher numbers.

choose whether or not to have a scan.

The the ultrasound scan on p1 looks like a photograph, but it cannot be one; it was made with sound not with light. You can also see that it is made up of small square picture elements, each one a particular shade of grey. These discrete picture elements are called pixels. The whole image is made up of 256×256 pixels.

The image was printed from information stored on a computer. This information was extracted from data obtained from the ultrasound scanner. The stored information is just a large array of numbers – one number for each pixel. The number for each pixel decides what shade of grey (or what colour) that pixel will be printed. In the image (top, right) you can see that darker regions correspond with larger numbers in the array.

The array is not an image, but is data for the image. It is easier for people to take in information presented as images rather than sets of numbers, a fact that has revolutionised much scientific work as people increasingly use images to look for important patterns in data. Furthermore, the image is made of numbers, so it can be modified by manipulating those numbers. The combination of visualisation and numerical processing is enormously effective for extracting and presenting information. Computer-generated and computer-manipulated images are increasingly changing art and entertainment, and influencing news and opinion. Thus an understanding of how this is done is well worth having.

Resolution

The resolution of an image is the smallest size of thing that can be distinguished. The image of the baby on p1 has limited resolution. No object smaller than one pixel can be seen. At 20 weeks' gestation a baby's limbs will, perhaps, measure 20 mm across. From the image on p1 you can see that the width of a pixel represents about 1 mm, so the resolution of the image is of the order of 1 mm.

In this situation the resolution of the image is very important. Suppose the baby's heart has a slight deformation: the resolution must be good enough to show it up.

The idea of resolution also applies to the grey levels in the scanned image. In this image there are just 256 possible different shades of grey recorded. A change in intensity smaller than this cannot be reproduced.

Resolution is a fundamental characteristic of all measuring systems. The resolution of any instrument is the smallest difference that is detectable. Look at a packet of cornflakes through a hand lens. Can you see dots of colour? Without the lens your eye does not resolve the dots so you see a continuous colour. Have you seen an Impressionist painting, for example by Seurat? His pictures are composed of dots of primary colours, which seen from a distance merge into a full range of hues. And of course the picture on your television set is also just a dance of coloured dots.

Key summary: ultrasound pulses

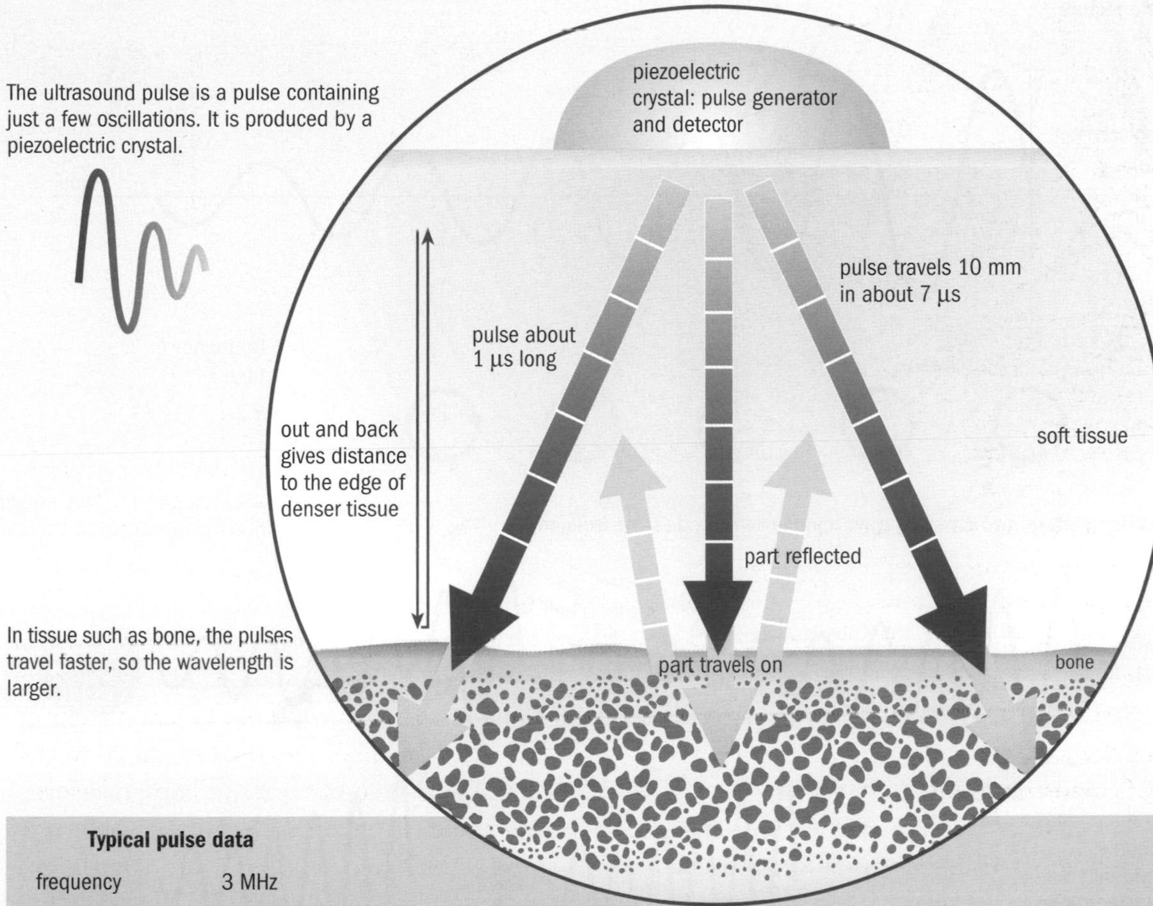

The ultrasound pulse is a pulse containing just a few oscillations. It is produced by a piezoelectric crystal.

piezoelectric crystal: pulse generator and detector

pulse travels 10 mm in about 7 μs

pulse about 1 μs long

out and back gives distance to the edge of denser tissue

soft tissue

part reflected

In tissue such as bone, the pulses travel faster, so the wavelength is larger.

part travels on

bone

Typical pulse data			
frequency	3 MHz		
speed in soft tissue	1500 metres per second (0.67 μs per mm)		
pulse length	about 1 μs (3 cycles)	typical scale of body parts	50–200 mm
wavelength	0.5 mm	typical out-and-back times	70–280 μs

An ultrasound scanner builds up an image from the strength and time delay of reflected pulses

Making a scan with ultrasound

An ultrasound scanner works like a depth-sounder, or "fish-finder", used by trawlermen. Short pulses of sound are sent out from the scanner, which is held in contact with the mother's belly. The pulses are reflected back where the density of the tissue changes. A baby, suspended in fluid in the womb, is quite a good reflector of sound. Bone is a very good reflector. Sound comes back to the scanner from the different parts of the baby, each part producing its own characteristic reflection. The farther away from the scanner a part of the baby is, the longer the delay between the pulse being sent and being reflected back.

The depths to be "sounded" are only about 100 mm, and ultrasound travels at high speed, taking about 70 μs to go 100 mm in soft tissue. Thus the delay times are short. The pulses must be only about 1 μs long if returning pulses are not to overlap outgoing ones. The time of an oscillation of the wave cannot possibly be longer than the time the pulse exists so the frequency must exceed 1 MHz, that is, one million oscillations per second. The highest frequency you can hear is probably about 20 kHz. So ultrasound is ultra-inaudible! The scanning system must be good at measuring very short time periods.

Ultrasound is generated by rapid electrical

Key summary: frequency, speed and wavelength

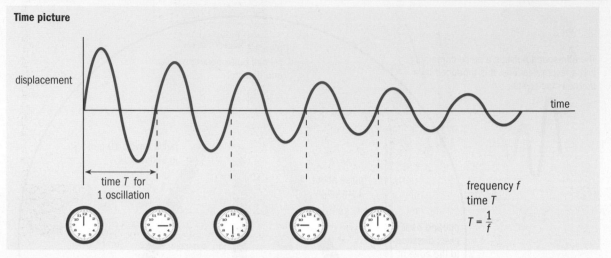

Time picture

displacement

time

time *T* for 1 oscillation

frequency *f*
time *T*

$T = \dfrac{1}{f}$

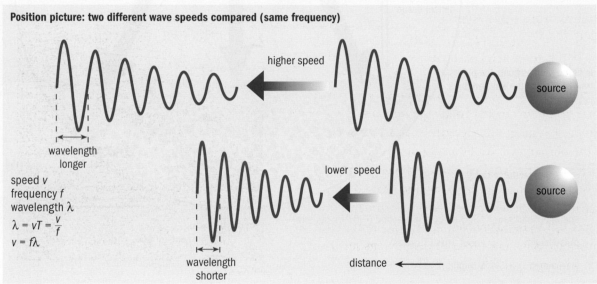

Position picture: two different wave speeds compared (same frequency)

higher speed

source

wavelength longer

speed *v*
frequency *f*
wavelength λ

$\lambda = vT = \dfrac{v}{f}$

$v = f\lambda$

wavelength shorter

lower speed

source

distance

Example

An ultrasound pulse lasts 1 μs, has a frequency of 10 MHz and travels at 2000 m s^{-1}. What can you calculate?

time *T* for one oscillation = $\dfrac{1}{f}$ = 0.1 μs

wavelength $\lambda = \dfrac{v}{f}$ = 0.22 mm

number of oscillations in one pulse = 10
length occupied by one pulse = 2 mm

Ultrasound pulses have a high frequency and a short wavelength

oscillations applied to a piezoelectric crystal. When the ultrasound pulse returns and vibrates the crystal, it produces a signal that is picked up by the scanner. You can buy an ultrasonic "tape measure" at most DIY shops. We will discuss more about this and other sensors in chapter 2.

The delay times of reflections tell the scanner the location of the denser tissue of the baby. The strengths of the reflections tell the scanner how dense the various tissues are. From the delay times and strengths of reflected pulses, a picture of the baby in the womb is built up. That's the principle. However, computer tomography – CT for short – does the big computing job of working out what is producing all the reflections. CT is widely used in medicine for many kinds of scan.

Henrietta Leavitt (left) discovered a way to measure distances to far-away galaxies. Jocelyn Bell Burnell (above) discovered pulsars.

Images from the universe

Until the development of radio astronomy in the 1940s, most news of the universe was "read" only through the narrow channel of visible light. Human eyes have evolved to see in just this narrow band of the spectrum. Today the universe is explored at many wavelengths that are invisible to the human eye. Radio astronomers observe at radio wavelengths; their early radio telescopes were adapted from radar instruments used in wartime. Other telescopes carry instruments to detect infrared, ultraviolet, X-ray and gamma-ray emissions. Each new development brings surprises. It is often particularly informative to look at the same object at different wavelengths; things invisible at one wavelength can be seen at another.

The image of Centaurus A in visible light (top, left) can be thought of as a photograph. One galaxy can be seen, but the second one that is colliding with it looks like a dusty band across the centre. In the infrared view, the shape and orientation of this second colliding galaxy can be seen. The data from the radio telescope have been used to plot contour lines of equal radio intensity, showing large lobes of radio activity well outside the regions picked up by the visible and infrared radiation images. Detailed study of the infrared image gives evidence of a black hole at the centre of the galaxy, which theory says should be expected to produce the radio emission seen.

How do we know that Centaurus A is 15 million light-years away? It was Henrietta

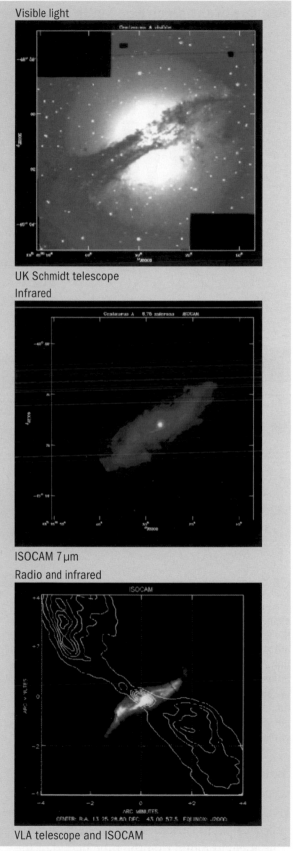

Visible light

UK Schmidt telescope

Infrared

ISOCAM 7 μm

Radio and infrared

VLA telescope and ISOCAM

Images of the radio source Centaurus A, one of the strongest radio sources in the sky. The images show what is thought to be a pair of colliding galaxies about 15 million light-years away.

Regions of the electromagnetic spectrum

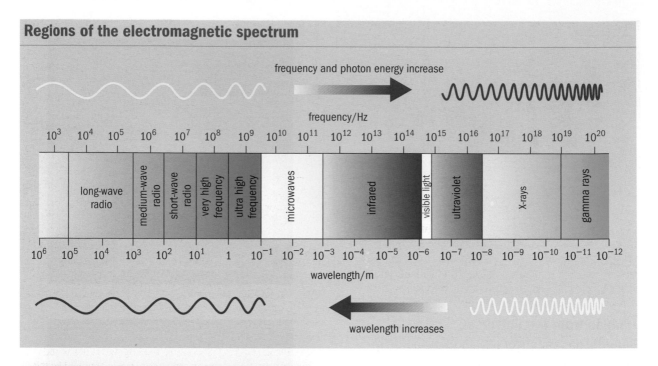

frequency and photon energy increase

frequency/Hz

10^3 10^4 10^5 10^6 10^7 10^8 10^9 10^{10} 10^{11} 10^{12} 10^{13} 10^{14} 10^{15} 10^{16} 10^{17} 10^{18} 10^{19} 10^{20}

long-wave radio | medium-wave radio | short-wave radio | very high frequency | ultra high frequency | microwaves | infrared | visible light | ultraviolet | X-rays | gamma rays

10^6 10^5 10^4 10^3 10^2 10^1 1 10^{-1} 10^{-2} 10^{-3} 10^{-4} 10^{-5} 10^{-6} 10^{-7} 10^{-8} 10^{-9} 10^{-10} 10^{-11} 10^{-12}

wavelength/m

wavelength increases

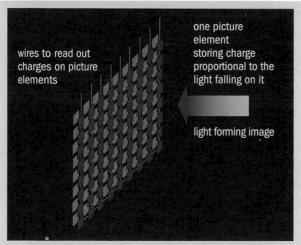

wires to read out charges on picture elements

one picture element storing charge proportional to the light falling on it

light forming image

The structure of a CCD. When an image is formed, each element stores a charge in proportion to the brightness of the light that has fallen on it. To record the image, charges are shunted in sequence from one element to the next, until they reach the edge where the value is read as a potential difference. Thus the whole image is recorded as a sequence of values of potential differences representing charges and therefore brightnesses. The picture becomes an array of numbers. The technology is continually evolving, and new kinds of digital light-sensitive devices are being developed.

Swan Leavitt (1868–1921), working at Harvard, who discovered a way to measure distances to far-away galaxies. The brightness of certain stars, called Cepheid variables, varies regularly with time. Leavitt found that the time period of the variation is related to the brightness of the star. So, given the time period, the brightness is known.

Given the brightness, compared with how bright the star looks, the distance can be found. Leavitt headed the Harvard Observatory Department of Photographic Photometry, setting standards for accurate work that were adopted internationally.

In the radio image of Centaurus A (p5) the resolution is low – little detail is visible. By contrast, the resolution of optical and infrared telescopes in space, outside the Earth's atmosphere, is now such that hundreds of planets have been detected circling nearby stars. The question of whether there is life elsewhere in the universe has become a live issue.

Astronomers Jocelyn Bell Burnell and Anthony Hewish were joking when they gave the name "little green men" to regularly pulsing signals they had noticed in the sky. Could they be from living things elsewhere in the universe? The thought is tempting but the reality was perhaps more exciting still. They had discovered pulsars – fast-spinning relics of stellar explosions, stars made of neutrons packed as densely as particles in an atomic nucleus. The radio telescope they used was crude and simple: just a large array of aerials joined together. Hammering in a huge number of metal stakes and stringing them with wire was hard work. While doing this they could not have anticipated the reward in store, but only hoped that a new channel of information might bring fresh news.

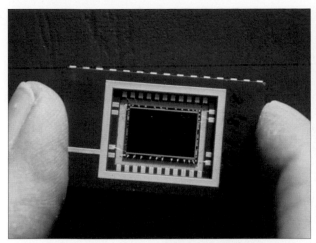

A typical CCD (charge-coupled device) chip. CCDs are used to capture images in cameras, webcams, TV cameras, portable videos and mobile telephones. A CCD can have more than 10 million pixels and can provide high-quality images in which individual pixels are far too small to see. Astronomers use even larger CCDs in telescopes, achieving very high resolution. Imaging technology is changing all the time and is likely to have been much improved by the time you read this.

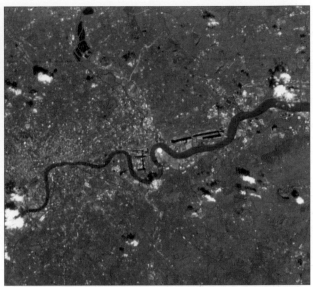

An infrared satellite image of London Docklands. Green vegetation reflects infrared radiation rather well, so areas covered with vegetation show up "bright" in infrared. In this image, the infrared signal controls the brightness of the red colour. But any colour could have been used. Pictures like this are called false-colour images.

A modern optical telescope uses a light-sensitive microchip instead of a photographic plate. One type is called a charge-coupled device (CCD). This is a screen covered by a million or more tiny silicon picture elements, each of which stores an electric charge when light falls on it. CCDs can be made that are sensitive to wavelengths outside the visible range. The image – a set of values of potential difference, one on each picture element – is read out element by element to be stored as a string of numbers. Back to numbers again.

Digital cameras, including the one in your mobile phone, webcams and camcorders all use CCDs or similar devices. Such microsensors (sensors built on a microchip) are becoming essential in modern instrumentation systems (this is discussed further in chapter 2).

Spy in the sky

International conflict has been an important source of scientific and technological development. For example, during the Cold War (about 1950 to 1990) the US and the then Soviet Union tried to detect each other's secret nuclear tests. Satellite imaging played a role in observing surface traces of underground testing.

School students made front-page news by tracking the first satellites. Satellite surveying is now big business. Digital maps covering much of the world are widely available on the internet.

An infrared satellite is able to track the growth of vegetation over the seasons, providing valuable ecological information. It can also detect areas in Europe where farmers are growing crops on land that is subsidised by the European Union to be left fallow. To the people who pay the subsidy this seems like detecting cheating. To the farmer it may feel like spying.

Satellites detecting infrared radiation are also used to monitor sea temperatures, for example changes in the Gulf Stream as it crosses the Atlantic keeping the climate of north western Europe relatively mild. Satellite images of ice cover in the Arctic and Antarctic, showing where ice has melted or glaciers have slipped into the sea, are important in the debate about the possible effects of global warming.

A satellite image of the Earth can easily have a resolution of 10 m or less. This means that you might be able to identify the building where you live, but would probably not be able to detect a parked car. You can estimate resolution if you enlarge an image until the pixels are visible. Then identify some object whose size you can guess, and see how wide it is measured in pixels.

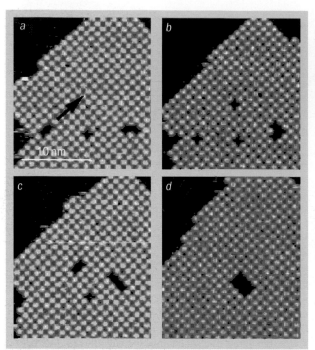

Moving molecules around. A layer of organic molecules has been deposited on a silver surface. There are several places where a molecule is missing from the regular pattern. A scanning tunnelling microscope has been used not only to image the layer of molecules, but also to push them around. Images (a) to (d) show successive stages as the molecules are moved to bring all of the empty sites together.

Scanning tunnelling microscope

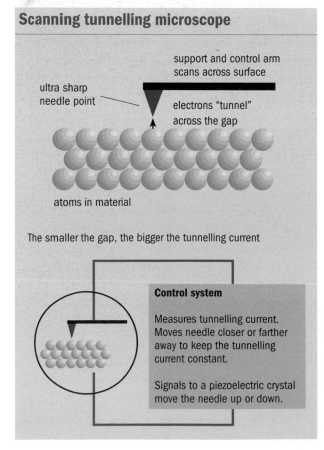

support and control arm scans across surface

ultra sharp needle point

electrons "tunnel" across the gap

atoms in material

The smaller the gap, the bigger the tunnelling current

Control system

Measures tunnelling current. Moves needle closer or farther away to keep the tunnelling current constant.

Signals to a piezoelectric crystal move the needle up or down.

Ernst Mach (1838–1916), an Austrian physicist and philosopher, is the namesake of the "Mach number" (also known as Mach speed). An aeroplane flying at Mach 2.0 is flying at twice the speed of sound in air in those conditions.

Seeing atoms

Austrian physicist and philosopher Ernst Mach hated what he called "fictive physics". Atoms, Mach said, are "things which can never be seen or touched and only exist in our imagination". Challenging physicists like Ludwig Boltzmann who thought of atoms as real, Mach would ask, "*Haben Sie einen gesehen?*" – "Have you seen one?" You should be aware that in Mach's time nobody had a good measurement of how big atoms are, or a good theory of what they might be made of. We believe that Mach did science a real service, even if in the end he was wrong. "You don't really know," he said. "It's only an idea." Science depends on imagination but also on doubt and criticism.

Mach died still not believing in atoms just as Max von Laue found the first visual evidence of their existence using X-rays, as Ernest Rutherford with Hans Geiger and Ernest Marsden discovered the nucleus within the atom and as Niels Bohr produced the first quantum theory of atoms.

Nowadays atoms can be "seen" and "touched". They can even be picked up and put down at will. The answer to Mach's question, "Have you seen one?" can be, "In a way, yes."

In chapter 5 you will learn about what can be understood from imaging atoms. For now, here is a set of images (top, left) showing molecules not only being seen but also being moved about.

There are several kinds of scanning microscope able to make images on the atomic scale, or near to it. One is the scanning tunnelling microscope.

A computer-generated image of a tornado developing. This storm never blew off a roof. It is not a picture of a cloud. It is not even an image of data about a cloud. It was made entirely by calculation, using a computer model of the development of storms.

This instrument has an ultrasharp needle point that is scanned in zigzags across the surface of a sample material. When the tip is close to the atoms on the surface, electrons can "tunnel" across the gap giving an electric current travelling between surface and microscope tip. The tip is moved up and down as necessary to keep the tunnelling current constant. A record of the movements of the tip is a record of the ups and downs of a material's surface on an atomic scale.

Does a scanning tunnelling microscope "see" atoms? What is actually observed is the electron current, not the atoms.

Images for the mind's eye

If seeing atoms seems unlikely, then seeing a theory must sound even more peculiar. But it isn't. Today, very often theories of things never seen, such as the way black holes might evolve, galaxies might form, or DNA molecules might fold up, are best thought of visually. A computer crunches numbers to simulate what might happen and then uses the numbers to make an image. The image is not a picture of the result: the image *is* the result (see the image at the top of the page of the development of a tornado).

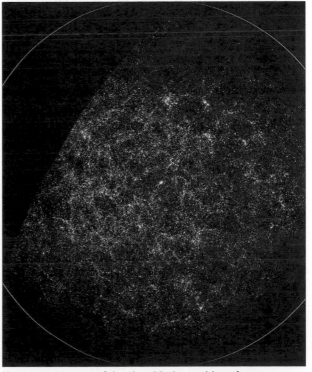

Here you see a part of the sky with the position of many galaxies plotted. The plot is designed to show whether or not the galaxies are evenly distributed in space. There is still debate about what these data say about the structure of the universe. Are the galaxies randomly distributed? Do they form structures like bubbles with voids in between? Data images like these are also images for the mind's eye.

Quick check

1. Ultrasound takes 70 μs to go 100 mm in soft tissue. Show that its speed is about 1400 m s^{-1}.

2. Suppose one part of a baby is 200 mm in front of another. Show that the difference in time between reflections from each part, from a pulse travelling at 70 μs per 100 mm, is about 280 μs. If the ultrasound pulses are 3.5 μs long show that the smallest detectable difference in distance is about 2.5 mm.

3. The ultrasound image of a baby on p1 is a square array of 256×256 pixels and the area imaged is about 250 mm across and 250 mm high. Show that the resolution of the image is about 1 mm in each direction.

4. Light takes 500 s to reach the Earth from the Sun. Light from Centaurus A (p5) takes 15 million years to reach the Earth. Show that Centaurus A is about 10^{12} times further away than the Sun at a distance of about 1.4×10^{20} km.

5. Centaurus A is 15 million light-years away, and a radio telescope can resolve structures of 1 s of arc across. Show that such a structure is 75 light-years across.

6. Show that there are
   (a) about 200 million atoms in the 20 mm width of a postage stamp;
   (b) about 1 million atoms in the 0.1 mm width of a pencil line;
   (c) 10 000 atoms in th 1 μm width of a pixel in a CCD.

Links to the *Advancing Physics* CD-ROM

Practise with these questions:
10S   Short answer *Speed, wavelength and frequency*
20E   Estimate *Large and small distances and times*
30E   Estimate *Making estimates about images*
30X   Explanation–exposition *Different kinds and uses of images*

Try out these activities:
10S   Software-based *Examples of images in physics*
20D   Demonstration *Electronic image capture*
30D   Demonstration *Distance measurement with ultrasound*

Look up these key terms in the A–Z:
Amplitude, frequency, wavelength and wave speed; atom; camera; CCD (charge coupled devices); electromagnetic spectrum; images; pixel; radians; resolution; ultrasound scanning

Go further for interest by looking at:
10I   Image *The variety of uses of scientific images*
20I   Image *Astronomical images: problems of noise and resolution*
70I   Image *Satellite images of towns in Europe*

Revise using the revision checklist

1.2 Information in images

Numbers in computers are stored as on or off values. "On" may be a high potential difference, "off" may be low. This is a binary system. The two values can be thought of as two digits, 1 and 0.

During the 1940s and 1950s, when computers were first coming into use, Claude Shannon, working at Bell Telephone Laboratories, devised a theory of how to measure information. One of his ideas was that the amount of information can be measured by the amount of storage it needs. If one pixel was simply either "bright" or "dark", only one memory location storing a 1 or a 0 would be needed. John Tukey named this one "bit" of information (contracting binary digit to bit). But often a pixel can have, say, 256 different shades of grey. One of 256 alternatives can be stored as a number from 0 to 255. That needs more memory space for each pixel. In fact it needs just eight memory locations   eight bits.

It is a curious fact that Shannon's theory measures information in a message without any regard to the meaning or content of the message. This enables the amount of information in a picture to be compared to that in a piece of text, a coded message or a table of numbers.

Bytes

One bit of information is one choice: 0 or 1. Eight bits, giving 256 possibilities, is a useful slice of memory. It is enough to give a code number to every key on a computer keyboard, including capital and lower-case letters, numbers and punctuation marks. Early in the history of computing, 8 bits was chosen as a unit of its own – one **byte**. Thus a 10 kbyte word-processed file that you have generated will contain around 10 thousand characters – and so about 1600 words (English words average about six characters each).

Actually, in computing, one kilobyte is 1024 bytes, not 1000 bytes, because 1024 is a simple power of two. Similarly one megabyte is a little more than 10^6 bytes.

Amount of information

The better the resolution of an image, the more information it contains. A digital colour camera image can have more than a million pixels, each

Claude Shannon (1916–2001). Shannon's mathematical theory of communication has proved important in linguistics, cryptography and in the design of communication systems.

Key summary: number of alternatives = 2^I

Key summary: bits and bytes (p12) shows that the number of alternative values that can be represented grows rapidly as the amount of memory used increases.

If 8 bits of information are used:

$$\text{number of alternatives } N = 2^8 = 256$$

In general, if the amount of information is I bits:

$$\text{number of alternatives } N = 2^I$$

If the number of bits increases by one, the number of alternatives doubles. Information is measured on a "plus" scale: the number of alternatives is measured on a "times" scale.

Plus scale	Times scale
amount of information increase by equal additions	number of alternatives increase in equal multiples
amount $= I$	number of alternatives $= 2^I$
amount $= \log_2 N$	number of alternatives $= N$

$\log_2$ **(number of alternatives) = information I = number of bits**

Key summary: bits and bytes

								Decimal value	Number of alternatives
8 bits = 1 byte				4 bits		2 bits	1 bit		
0	0	0	0	0	0	0	0	0	
0	0	0	0	0	0	0	1	1	
0	0	0	0	0	0	1	0	2	$2^1 = 2$
0	0	0	0	0	0	1	1	3	
0	0	0	0	0	1	0	0	4	$2^2 = 4$
0	0	0	0	0	1	0	1	5	
0	0	0	0	0	1	1	0	6	
0	0	0	0	0	1	1	1	7	
0	0	0	0	1	0	0	0	8	$2^3 = 8$
. . .								. . .	
0	0	0	0	1	1	1	1	15	
0	0	0	1	0	0	0	0	16	$2^4 = 16$
. . .								. . .	
0	0	0	1	1	1	1	1	31	
0	0	1	0	0	0	0	0	32	$2^5 = 32$
. . .								. . .	
0	0	1	1	1	1	1	1	63	
0	1	0	0	0	0	0	0	64	$2^6 = 64$
. . .								. . .	
0	1	1	1	1	1	1	1	127	
1	0	0	0	0	0	0	0	128	$2^7 = 128$
. . .								. . .	
1	1	1	1	1	1	1	1	255	
1	0	0	0	0	0	0	0	0	256	$2^8 = 256$

One 8-bit byte stores $2^8 = 256$ alternatives

using 3 bytes (24 bits) for colour information about each pixel. So the image may need 3 Mbyte of storage. Storing images on a computer rapidly uses up the storage capacity, even if it has 100 Gbyte or more. And processing images places big demands on memory while the image is being worked on, because many operations have to be performed.

Computers claim to offer "millions of colours". How few bits are needed to represent so many colours? The number *n* of bits must be big enough

Key summary: logarithmic scales

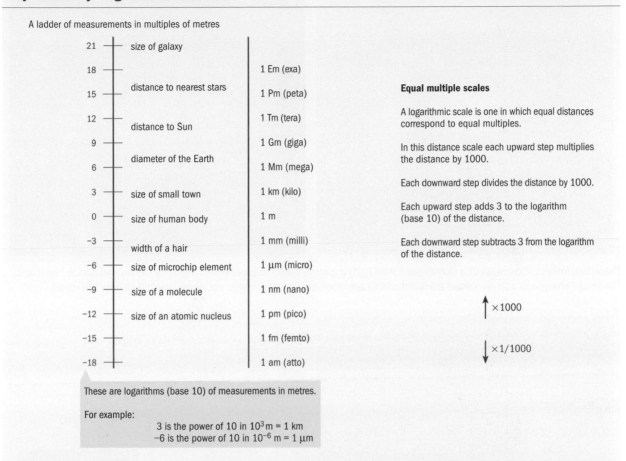

A ladder of measurements in multiples of metres

21	size of galaxy	
18		1 Em (exa)
15	distance to nearest stars	1 Pm (peta)
12	distance to Sun	1 Tm (tera)
9		1 Gm (giga)
6	diameter of the Earth	1 Mm (mega)
3	size of small town	1 km (kilo)
0	size of human body	1 m
−3	width of a hair	1 mm (milli)
−6	size of microchip element	1 μm (micro)
−9	size of a molecule	1 nm (nano)
−12	size of an atomic nucleus	1 pm (pico)
−15		1 fm (femto)
−18		1 am (atto)

These are logarithms (base 10) of measurements in metres.

For example:

3 is the power of 10 in 10^3 m = 1 km
−6 is the power of 10 in 10^{-6} m = 1 μm

Equal multiple scales

A logarithmic scale is one in which equal distances correspond to equal multiples.

In this distance scale each upward step multiplies the distance by 1000.

Each downward step divides the distance by 1000.

Each upward step adds 3 to the logarithm (base 10) of the distance.

Each downward step subtracts 3 from the logarithm of the distance.

↑ ×1000

↓ ×1/1000

Using a logarithmic scale, distances of many orders of magnitude can be shown

that 2^n is more than a million. Experiment with different values on your calculator. The number of bits (n) needed is just 20. So if each pixel has to store one of at least a million possibilities, 2–3 bytes are needed for each (2 bytes are 16 bits; 3 bytes are 24 bits). The number 20 is approximately the logarithm to the base two of one million.

You get information from a message because it communicates one possible alternative out of many. Information engineers use logarithms to measure amounts of information. Sound engineers use logarithms to measure sound intensity in decibels. Astronomers use logarithms to measure the brightness of stars in magnitudes. Logarithms are often a good way to make plots of quantities that span several orders of magnitude. You will use logarithms a lot. They make some kinds of thinking much easier, once you have got your head round how logarithms work.

Logarithms, a way to use small numbers to think about big numbers by turning multiplying into adding, are a clever and elegant idea. They were invented long ago by John Napier (1550–1617), an independent-minded Scot fascinated by the idea of doing all calculations by machine. His logarithms made the tedious astronomical calculations of the time much easier — the only calculators then were people. Napier believed no less strongly in astrology and divination, was fiercely involved in the religious disputes of the time, and designed a primitive tank he hoped would be used in the war against Philip II of Spain.

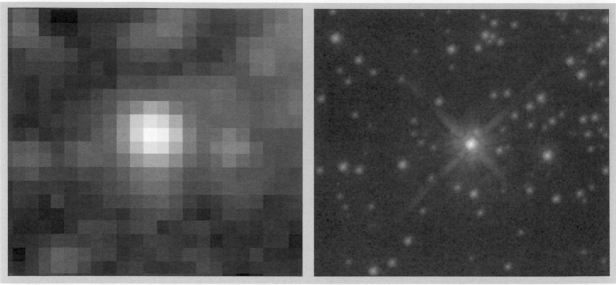

These two images show stars in a cluster seen from (left) a ground-based telescope and (right) by the Hubble Space Telescope. The image sharpness and number of stars detectable are much better in the image obtained from the Hubble telescope.

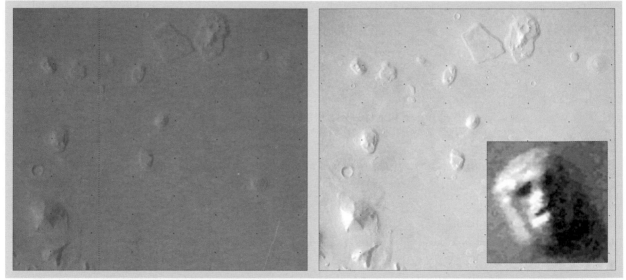

Left: an unprocessed image of the surface of Mars taken by the *Viking 1* Orbiter. Right: the same image with noise removed and contrast enhanced. Inset: there has been controversy over the face-like markings in one part of the image. Are they real, or have they been introduced as a result of image processing?

Image processing

When images were mainly photographs, only a limited amount could be done to enhance them. But stored as numbers, images can be transformed in many ways – changing colours, removing and adding pieces. You can no longer be sure that a photograph in a newspaper tells it as it was.

Noise in images can appear as a random speckle. It can be reduced by smoothing. One way to remove noise is to replace the value of each pixel with the median (the middle value in order) of its value and those around it. Another way is to replace each value with the arithmetic mean of it and its neighbours. Using the mean also rounds off sharp corners and blurs sharp edges. In general, averaging is a good way of removing random or rapid variations in all sorts of experiments.

Edges are important clues if you are looking for objects in an image. To enhance edges, instead of averaging a pixel with its neighbours, the average value of its neighbours is subtracted from each pixel value. This removes uniform areas of brightness and picks out just the places where the gradient of the brightness changes abruptly – at the edges.

Key summary: image processing

Smoothing sharp edges

rule

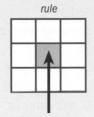

replace each pixel by the mean of
its value and those of its neighbours

effect if there is an edge

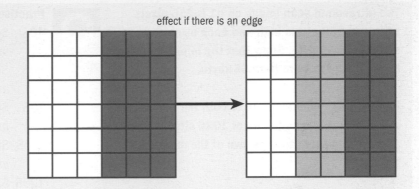

Removing noise

rule

replace each pixel by the median of
its value and those of its neighbours

effect if there is noise

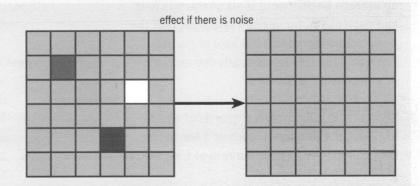

Finding edges

rule

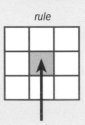

	−1	
−1	+4	−1
	−1	

subtract the N, S, E and W neighbours
from 4 times the value of each pixel.

effect if there is an edge

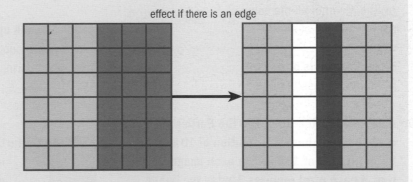

This is called the Laplace rule. It
detects changes in the gradient of
brightness.

effect if there is no edge

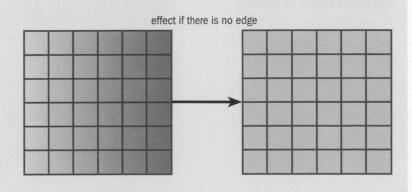

Images can be manipulated and improved by modifying pixel values

Quick check

1. The ultrasound scan image on p1 is 256 pixels wide and 256 pixels high and each pixel is specified by 1 byte. Show that the information in the image is more than 65 kbyte.

2. Explain why the number 10 is both the number of bits needed to code one of 1024 alternatives and the logarithm to base two of the number 1024.

3. A good sized book has about 100 000 words. If an average English word is six characters long and each character needs 1 byte show that the book can be stored in 600 kbyte of memory. Show that the text takes nearly five seconds to be transmitted at 1 Mbit s^{-1}.

4. Show that it takes nearly a quarter of an hour to transmit 100 images, each of 1 Mbyte on a channel with a transmission rate of 1 Mbit s^{-1}.

5. In this array, replace the "pixel" in the middle with
   (a) the mean of all the values;
   (b) the median of all the values.
   Which system is better for eliminating possible noise?

100	100	100
100	200	100
100	100	100

6. A satellite system to image the Earth's surface is designed to have a resolution of 10 m and to cover an area of 100 km^2 in each image. Show that if each pixel requires 3 bytes the image requires 3 Mbyte of memory.

Links to the *Advancing Physics* CD-ROM

Practise with these questions:

40S Short answer *Smoothing pixels using mean or median*

60S Short answer *A scanning electron microscope image of Velcro*

70S Short answer *Image processing by brightness and contrast control*

110S Short answer *Bits and bytes in images*

120S Short answer *Logarithms and powers*

130C Comprehension *X-ray image of the Kepler supernova remnant*

Try out these activities:

50S Software-based *Image processing: the surface of Mercury*

60S Software-based *Image enhancing: volcanoes on Io*

70S Software-based *Medical uses of X-ray images*

80S Software-based *Medical uses of ultrasound images*

Look up these key terms in the A–Z:
Amount and rate of transmission of information; average; bits and bytes; image processing; indices; logarithmic scales; noise; prefixes

Go further for interest by looking at:

30I Image *X-ray images in medical physics*

40I Image *Ultrasound images in medical physics*

50I Image *Magnetic resonance imaging*

60I Image *Gamma rays detecting abnormalities*

Revise using the revision checklist and:

50O OHT *Bits and bytes*

60O OHT *"Plus" and "times" scales of information*

1.3 With your own eyes

Images and visualisation are important because eyes, and the brain of which they are an extension, are such clever instruments.

The eye is like a camcorder. But it is like a smart camcorder with intelligent computerised control of direction, focus and resolution, and with on-board chips processing signals before sending them on to the computer.

You can learn several surprising things about your intelligent eye by some simple tests.

How to see in the dark

Nearer the edges of the retina, rods and cones are less tightly packed, so the resolution of off-centre vision is poor. To see clearly you have to move your eyes so that the desired part of the image falls on the fovea, which is densely packed with cones.

A good way to see faint stars at night is not to look directly at them. If you look just to one side, the star will look brighter. This is because the rods in the retina are sensitive to low levels of light, but there are none of these rods in the fovea.

Your eyes

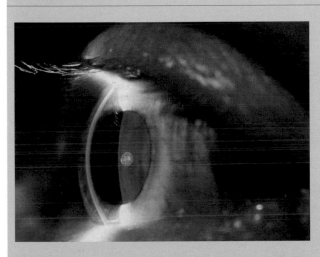

Left: light carrying information from the outside world enters the eye through its transparent curved front surface, the cornea. Much of the bending of light needed to form an image happens here. A lens inside the eye adjusts the focus to produce a sharp image on the retina.

Below: inside the eye is an adjustable lens that adds to the bending of light by the curved surface of the cornea. Muscles attached to the lens can stretch the lens, making it less curved, helping to adjust the eye to focus at a given distance.

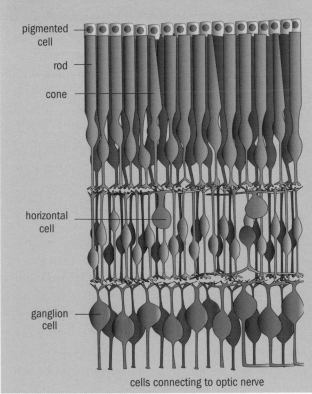

cells connecting to optic nerve

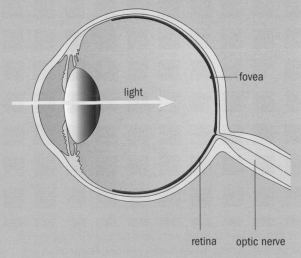

Left: the retina of the eye is built of light-sensitive elements that emit signals that go to the brain. Cones detect colours and work best in bright light. Rods are sensitive to dim light but do not distinguish colours. There are more than 100 million rods in the eye. Rods greatly outnumber cones, except in one special central region, the fovea, where there are only cones. This is where the eye detects fine detail.

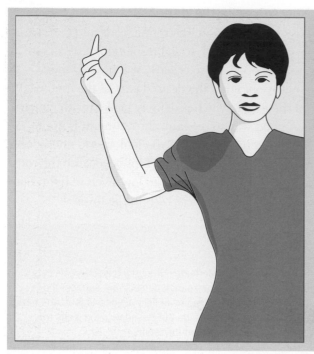

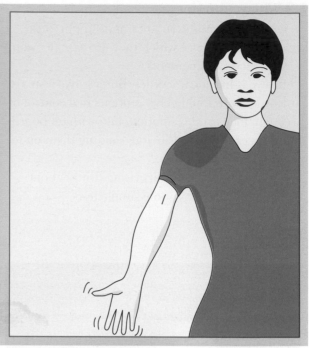

Look straight ahead and hold your hand out to one side, but don't look at it. How many fingers can you count? You probably can't count them. Look straight ahead and hold your hand by your side. You probably can't see it. If you waggle your hand you should see the movement, but you probably can't see your hand. Your brain just detects "something moving".

This grid of black squares looks as if it has greyish patches at the crossing points. Artists such as Bridget Riley exploit this kind of optical illusion in their work.

Your retina "thinks"

The rods and cones in the retina are connected to those near them. The connections are inhibitory; a lot of light on one rod or cone turns down the sensitivity of its neighbours. The rods and cones "seeing" the crossing points of the grid have their

sensitivity turned down because many of those near them are also brightly lit. So you see greyish patches at those places.

This system of connections works like the sharpening of an image (see pp14 and 15). It makes the eye into a good detector of sharp edges. Indeed, the eye exaggerates the contrast of bright and dark at an edge. This clever processing, done entirely in the retina, helps you to see the world as made out of well-defined objects with definite boundaries. The grey patches that you see in the grid are your retinal edge-detectors working overtime.

Other effects like this have been exploited by painter Bridget Riley. Some of her paintings look as if they are moving.

Shaping light to make images

Optics must have started in ancient times when people found glass fused from sand in spent fires. The first telescopes, in the hands of Galileo Galilei (1564–1642), helped shake the foundations of the then current view of the universe. Some people refused to believe that Galileo could see sunspots, the phases of Venus and the moons of Jupiter. Nor did they like the idea that the Earth might be just one body in the heavens among others. Disturbing

Waves spread out in all directions from a point source like the circular ripples produced by a stone thrown in a pond.

thoughts arose: might not other heavenly bodies support life? – a question not answered even today. And worse, what religion would God have given such living beings? The Italian monk Giordano Bruno was burnt at the stake for this heretical idea.

Images are made using lenses or mirrors. There are several ways of thinking about how lenses and mirrors shape light to make images. We have decided to use mainly the wave point of view, but will link it to the ray point of view, which you will find in many other books. There is value in more than one point of view:

- one explanation helps with some problems; another helps more with others;
- there really are several different ways of thinking about light (see chapters 6 and 7).

In the ray point of view, light is tracked along rays that go straight until the light is reflected at a surface or is bent as it passes from one substance into another. Rays from a point source spread out in all directions, getting farther and farther apart. From the wave point of view, waves spread out in all directions from a point source like the circular ripples produced by a stone thrown in a pond.

The connection between the two points of view is that a ray is a line pointing along the direction of motion of the wavefront. Rays are always at right angles to the wavefronts at the point where they

Anton van Leeuwenhoek (1632–1723), working in Delft in Holland, saw with his single-lens microscopes (which could magnify up to 300 times) a whole new world of the very small. The microscope in the photograph is of a kind developed from those made by van Leeuwenhoek.

cross. Rays are convenient construction lines, not real light paths, although the beam from a laser does stay quite close to a single straight line.

What a lens does

A lens used as a "burning glass" brings light across the whole surface of the lens down into a tiny image of the Sun, at the point called the focus (named after the Latin word for hearth). In ray language, the burning glass works by bending rays, bringing them from being parallel to coming together at a point. In wave terms, the lens works by altering the curvature of the wave, changing it into spherical ripples that converge on a focus.

The lens makes a wave more curved by slowing down the middle of a wave more than its edges. The part of the wave going through the thick glass

Key summary: rays versus waves

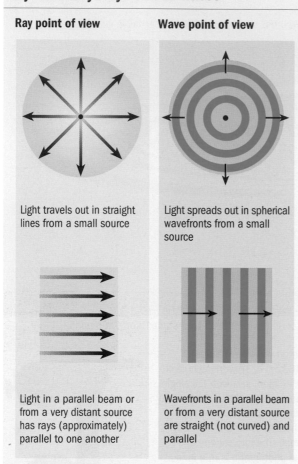

Ray point of view

Light travels out in straight lines from a small source

Light in a parallel beam or from a very distant source has rays (approximately) parallel to one another

Wave point of view

Light spreads out in spherical wavefronts from a small source

Wavefronts in a parallel beam or from a very distant source are straight (not curved) and parallel

Ray and wave points of view show the same thing but in different ways

Key summary: burning glass

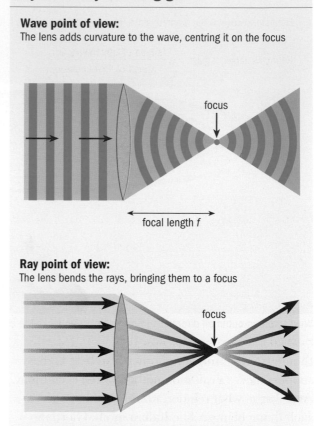

Wave point of view:
The lens adds curvature to the wave, centring it on the focus

focus

focal length *f*

Ray point of view:
The lens bends the rays, bringing them to a focus

focus

A lens changes the curvature of a wavefront: this is equivalent to bending the rays

in the middle of the lens is delayed more than the part of the wave going through the thinner glass near the edges of the lens, so the middle portion of the wave gets left behind a little.

The distance from lens to focus is called the focal length *f*. A powerful lens has a small focal length. A typical camera lens has a focal length of about 50 mm. A spectacle lens for a long-sighted person may have a focal length of 250 mm or so.

A lens changes how curved the waves are. If the waves are not curved at all when they get to the lens, they become spheres centred on the focal point after going through the lens. The radius of the spherical wavefronts just after passing through the lens is *f*, the focal length. The curvature of a sphere of radius *r* is $1/r$, so the lens adds a curvature of $1/f$ to the wavefronts. The power of a lens (in dioptres) is $1/f$, i.e. the curvature that

it adds to the wavefronts (*f* is in metres). Thus, a camera lens with focal length of 50 mm has a power of $1/(50 \times 10^{-3}) = 20$ dioptre.

A spectacle lens with focal length of 0.5 m has a power of $1/0.5 = 2$ dioptre. There are many other examples in physics of such pairs of reciprocal quantities, where $1/a = b$ and $1/b = a$ (see resistance and conductance in chapter 2).

Photographers and astronomers mostly work with focal lengths. Opticians work with lens powers in dioptres. You take your choice, but it is handy to be able to speak both languages. The big advantage of lens powers is that they add up: the power of two thin lenses put in contact is the sum of the powers of each individual lens.

Refractive index

Colourless glass can make a colourful sculpture. Where does the colour come from? Reflection and refraction of light play a part. So do transmission,

Key summary: defining refractive index

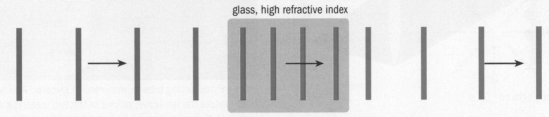

Light slows down when it enters glass, and speeds up again as it leaves. The wavefronts are closer together in glass than in air.

Refractive index of glass $n = \dfrac{\text{speed of light in vacuum}}{\text{speed of light in glass}}$

glass, high refractive index

Value for one type of glass: $n = 1.5$, so speed of light in this glass $= \dfrac{3 \times 10^8 \text{ m s}^{-1}}{1.5} = 2 \times 10^8 \text{ m s}^{-1}$

Light travels slower in glass than in air or in a vacuum

absorption and dispersion. It's a complicated business and artists working with glass have learned from experience how to achieve the effects they desire. Physicists try to explain these effects.

It all comes down to a question of the speed of light. Light travels fastest in a vacuum, at almost $3 \times 10^8 \text{ m s}^{-1}$. It travels only slightly slower in air (because air is mostly a vacuum). But when light travels from air into glass, it slows down a lot. This is because light interacts with the electrons in the atoms of the glass. From the refractive index n of the glass we can work out how much the light slows down. This depends on the composition of the glass – there isn't just one type of glass. Artists have traditionally used lead crystal glass for their finest work; this has a high refractive index and so has the biggest effect on light.

A lens works because light travels more slowly in the glass or plastic of the lens than it does in air. The ratio of the two speeds is called the refractive index of the transparent material.

$$\text{refractive index} = \frac{\text{speed of light in a vacuum}}{\text{speed of light in a material}}$$

Glass and plastic used for spectacle lenses commonly have refractive indices in the 1.4 to 1.6 range. The larger the refractive index the slower light travels inside the lens. To add the same curvature to a wavefront, a lens of high refractive index can be less curved than one of low refractive index. This means that the lens can be made

Glass sculpture "Space" (1980) by Czech artist Pavel Hlava. Light is reflected and refracted in the interior of the sculpture.

thinner. Thin, lightweight spectacles aim to use materials of high refractive index.

Now you see it...

Here's one way to find the refractive index of a piece of glass. Immerse it in some clear oil. If the oil and glass have the same refractive index, the glass will be virtually invisible.

How it works: when the two refractive indices match, light doesn't change speed as it passes from oil to glass. There is no reflection or refraction, so we can't see the glass. Forensic scientists use this

Key summary: how a lens makes an image

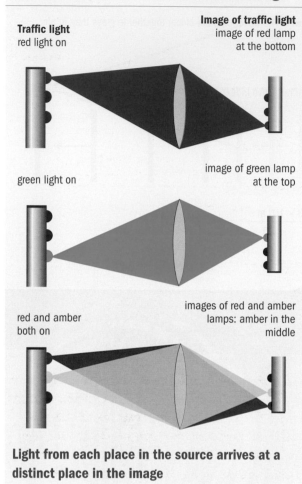

Traffic light
red light on

Image of traffic light
image of red lamp
at the bottom

green light on

image of green lamp
at the top

red and amber
both on

images of red and amber
lamps: amber in the
middle

Light from each place in the source arrives at a distinct place in the image

Here a Pyrex boiling tube is immersed in glycerol. At a specific temperature the refractive indices of the two materials are identical. Not only does the boiling tube look invisible in the beaker, it casts no shadow. Mineralogists keep sets of calibrated oils in their laboratories that cover a range of refractive indices. They can identify transparent crushed samples of minerals under a microscope by finding which oil makes the fragments vanish when a drop of the oil is put on the mineral. If different oils are mixed, an average of the refractive indices of the oils is taken, this can provide greater precision in identifying a material.

not enough power to add the needed curvature to the waves to bring them to a focus on the retina. The answer is to add more power. If you use a magnifying glass in front of your eye you can have the image on your retina big and in focus.

You can't get closer to a galaxy to see it better, so a telescope uses a lens or a mirror to bring an image of a star closer to your eye. You then look at this image with a magnifying glass (the eyepiece).

The rule for how lenses shape light

The rule for how a lens shapes light is simple. A lens changes the curvature of the wavefronts by a fixed amount, $1/f$, or:

curvature of waves going out equals curvature of waves coming in plus curvature added by the lens.

This can be translated into a rule for distances of image and object from the lens, and the focal length, which is a more natural way of thinking from the ray point of view. The rule is:

$$\frac{1}{v} = \frac{1}{u} + \frac{1}{f}$$

where u is the distance of the source from the lens, v is the distance of the image of the source from the lens and f is the focal length.

If you read the rule as a formula linking

technique to identify fragments of glass found at the scene of a crime.

The making of images

A lens makes an image "like a little picture" because light goes through the lens from every part of the object in front of the lens. Light from a particular point of the object comes together again behind the lens at a particular point in the image. The general direction of travel of the light is not altered by the lens – only the curvature of the waves – points high up on the object appear low down on the image, and points on the right of the object appear on the left of the image.

To see detail, the important thing is the size of the image on your retina. Bring your eye quickly closer to this page. The objects on the page zoom up in size as the image on the back of your eye gets bigger. But they get fuzzy too – your eye has

Key summary: lenses change wave curvature

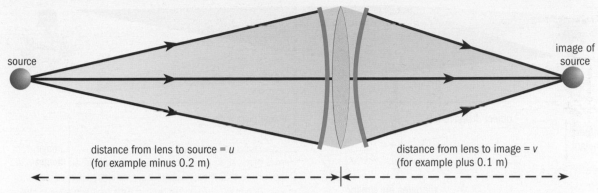

source

image of source

distance from lens to source = u
(for example minus 0.2 m)

distance from lens to image = v
(for example plus 0.1 m)

curvature of wavefront after leaving the lens = $\dfrac{1}{v} = \dfrac{1}{0.1} = 10$

curvature of wavefront on reaching the lens = $\dfrac{1}{u} = \dfrac{1}{-0.2} = -5$

curvature added by lens = power of lens = curvature after − curvature before = $\dfrac{1}{v} - \dfrac{1}{u} = 10 - (-5) = 15$ dioptre

power of lens $\dfrac{1}{f} = 15$ dioptre; focal length of lens $f = \dfrac{1}{\text{power}} = \dfrac{1}{15} = 0.067$ m = 67 mm

A lens adds curvature 1/f to waves passing through it

reciprocals of distances, the lens rule looks hard to understand. If you read it as a rule that says that a lens changes the curvature of wave fronts by a fixed amount, its meaning becomes clearer.

The rule for the signs of lens distances is like the rule for signs when drawing graphs:
- measure from the lens;
- count distances to the right as positive;
- count distances to the left as negative.

Because these rules are like cartesian graph rules, they are referred to as the cartesian sign convention.

If the source is far away, the wavefronts are flat and the lens makes them curved. If you bring the source up to the focal point of the lens, the lens adds curvature to the negatively curved wavefronts from the source, making them flat. In between, both incoming and outgoing wavefronts are curved, and the difference between their curvatures is the amount $1/f$ added by the lens.

Light takes the same time to travel all paths through a lens

There is another remarkable way of looking at what light does as it goes through a lens. It is this: light clocks up exactly the same time on all paths from a point on a source to the corresponding point on the image. Why? The angled outer paths

are obviously longer than ones through the centre, so why doesn't light take longer to travel the longer path? The reason that the shorter path through the centre includes more glass, in which the light travels slower. The longer path through the edge includes more air, in which the light travels faster. The shape of the lens is designed so that all paths take the same time to travel regardless of the thickness of glass the light has to pass through.

It has to be like this. After all, the lens just adds curvature to the wave and manages to bring the whole wave together at the image, with all parts of the wavefront still in step. Keeping the wavefront together is making all parts of it take the same time to travel. Chapter 7 shows how this idea can be taken further still into quantum behaviour.

Keeping things simple

The theory of lenses that we have given here has been simplified. It works only if the width of the lens is small compared to its focal length, and if rays pass at small angles to the axis of the lens. For more complicated lenses, such as zoom lens on a camera, professionals use computer ray-tracing, or treat the lens as transforming source into image. Computer ray-tracing is also used to generate realistic-looking artificial scenes.

Key summary: magnification by a lens

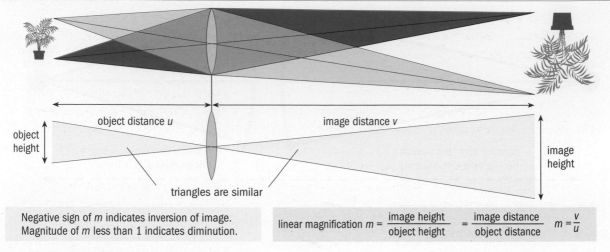

object distance u

image distance v

object height

image height

triangles are similar

Negative sign of m indicates inversion of image. Magnitude of m less than 1 indicates diminution.

linear magnification $m = \dfrac{\text{image height}}{\text{object height}} = \dfrac{\text{image distance}}{\text{object distance}}$ $m = \dfrac{v}{u}$

The linear magnification of a lens is m = v/u

Key summary: lenses add constant curvature 1/f

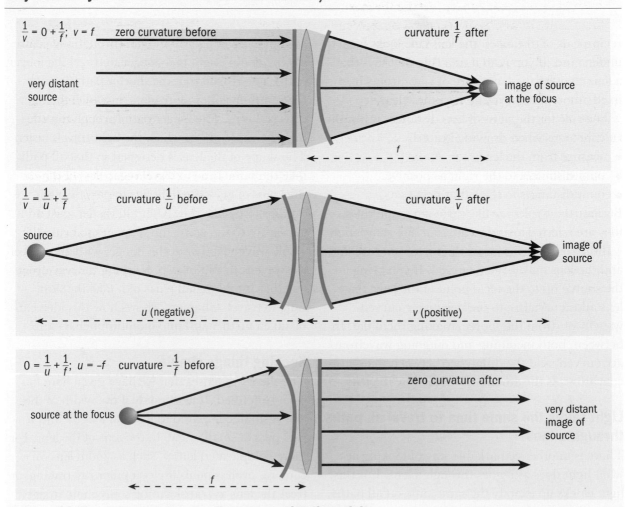

$\dfrac{1}{v} = 0 + \dfrac{1}{f}$; $v = f$ zero curvature before

curvature $\dfrac{1}{f}$ after

very distant source

image of source at the focus

f

$\dfrac{1}{v} = \dfrac{1}{u} + \dfrac{1}{f}$ curvature $\dfrac{1}{u}$ before

curvature $\dfrac{1}{v}$ after

source

image of source

u (negative)

v (positive)

$0 = \dfrac{1}{u} + \dfrac{1}{f}$; $u = -f$ curvature $-\dfrac{1}{f}$ before

zero curvature after

source at the focus

very distant image of source

f

A lens adds the same curvature to all waves passing through it

Quick check

1. A 100 dioptre lens in a CD player focuses a parallel beam of laser light on the disc. Show that the disc is 10 mm from the lens.

2. An astronomical telescope for home use has a mirror with a focal length of 800 mm. Show that the power of the mirror is 1.25 dioptre.

3. An elderly person's bifocal spectacles have a long-distance viewing area with a power of −0.2 dioptre and a reading area with a power of 3.3 dioptre. Show that the focal lengths of the two areas are respectively 5.0 m and 300 mm. Explain why the glass is most strongly curved in the reading area of the spectacles.

4. A camera with a 50 mm focal length lens is focused on a person's face 1 m from the lens. Show that the film will be about 53 mm behind the lens. (Remember to count the object distance as a negative value.)

5. With a zoom lens you can vary the power of a lens. Show that a lens power of about 24 dioptre is required to focus a flower that is 250 mm in front of the lens on a piece of film that is 50 mm behind the lens.

6. A lamp 400 mm from a lens is in focus on a screen 400 mm from the lens on the other side of the lens. Show that the lens has a power of 5 dioptre and a focal length of 200 mm. Explain why the lamp must be moved nearer to the lens to project its image further away.

Links to the *Advancing Physics* CD-ROM

Practise with these questions:

140S Short answer *Response of the human eye to differences in brightness*

150S Short answer *Cameras and eyes*

160C Comprehension *Satellite imaging*

50D   Data handling *Representing experimental uncertainties graphically 1*

60D   Data handling *Representing experimental uncertainties graphically 2*

Try out these activities:

130D Demonstration *Grey step: edge enhancement in the retina*

140E Experiment *The intelligent eye*

Look up these key terms in the A–Z:

Eye; focal length, power and magnification of a lens; lens; refraction

Go further for interest by looking at:

20T   Text to read *A collection of visual illusions*

Revise using the revision checklist and:

1600 OHT *Rays and waves spreading*

1700 OHT *Rays and waves focused*

1800 OHT *Formation of an image*

1900 OHT *Action of a lens on a wavefront*

2000 OHT *Where object and image are to be found*

Summary check-up

Waves ✓

- A wave of frequency f travelling at speed v has wavelength λ given by $v = f\lambda$

Images and information ✓

- Images can be formed with many kinds of signals, including ultrasound and all parts of the electromagnetic spectrum
- Images can be recorded electronically by microsensors; an example is the charge-coupled device (CCD)
- Images on the atomic scale can be recorded by scanning methods; an example is the scanning tunnelling microscope
- Images can be stored as an array of pixels, each defined by a number
- Images can be smoothed by suitable averaging; images can also be sharpened by locating edges

Logarithms ✓

- Quantities that cover a large range of values can usefully be displayed on a logarithmic scale. Prefixes for scientific units (e.g. micro, milli, kilo, mega) are chosen at multiples of 1000.
- 1 bit of information is two choices (0 or 1); 1 byte of information contains 8 bit (256 alternatives); amount of information I provides $N = 2^I$ alternatives

Eye and lenses ✓

- The eye is like an intelligent video camera, sending out a stream of processed signals. It detects edges and movement.
- A converging lens adds a constant curvature to light falling on it. The curvature added is the power of the lens. Power of lens $= \dfrac{1}{f}$.
- Image and object distances $\dfrac{1}{v} = \dfrac{1}{u} + \dfrac{1}{f}$ (cartesian sign convention)
- The linear magnification of a lens $m = \dfrac{v}{u}$
- Light takes the same time to travel on all paths from a point on the source to the image via the lens (or mirror)
- refractive index $= \dfrac{\text{speed of light in a vacuum}}{\text{speed of light in a material}}$

Questions

1. **A digital camera has a lens of focal length 50 mm and produces an image of 1280×960 pixels. A human face (270×200 mm) that is 1.5 m from the lens fills the picture area of the light-sensitive chip.**
   (a) What is the power of the lens in dioptres?
   (b) Explain why when the face is in focus the light-sensitive surface will be only a little more than 50 mm from the lens.
   (c) How many pixels make up the image?
   (d) Estimate the dimensions of the picture area of the light-sensitive chip in the camera.
   (e) Estimate the dimensions of one picture element in the light-sensitive chip.
   (f) Estimate the scale of features on the face that can be resolved in the picture. Could a person's eyelashes be resolved?
   (g) If the camera stores three colours for each pixel, each with 256 levels of intensity, how big will a picture file be?

2. **Make a logarithmic scale of the size (i.e. linear dimensions) of living organisms from viruses to whales. Make each point on the scale 10 times as large as the one before.**
   (a) Try to place a representative organism at each point on the scale.
   (b) Suggest reasons why there are no organisms one-tenth of the size of viruses.
   (c) Suggest a reason why there are no organisms 10 times as large as a whale.
   (d) Show how you can easily add scales for the volume and mass of organisms (you may suppose their density is close to that of water, $1000 \, \text{kg m}^{-3}$).

3. **Spectacles for people with long sight use converging lenses.**
   (a) Draw a diagram to show how wearing these spectacles helps a person to focus sharply on the print in a book close to them.
   (b) Describe how to measure approximately the power of the lenses in such spectacles.
   (c) If the eye has a length of about 25 mm from front to back, estimate the power of the combination of cornea and eye lens.

   (d) A combination of two thin lenses in contact has a power approximately equal to the sum of the powers of the two lenses. Suggest an explanation of this rule. Translate it into an expression for the focal length F of the combination in terms of the focal lengths f_1 and f_2 of the two lenses.

4. **Give your own example of a scientific imaging system (not necessarily using light) producing a computer-storable image.**
   (a) Explain how it works and how the image is made.
   (b) What limits the resolution of the image?
   (c) Suggest one advantage and one disadvantage of the system over others.

5. **A balloon filled with carbon dioxide, in which sound travels more slowly than in air, can be used to focus sound waves, like a lens used to focus light.**
   (a) Sketch a diagram of sound waves from a distant source passing through the balloon, showing how they come to a focus.
   (b) Explain how the effect depends on sound travelling more slowly in carbon dioxide than in air.

6. **This is an image of atoms forming a grain boundary. The image is spoiled by random noise, which gives the image a speckled appearance.**

   (a) What, approximately, must be the resolution of the image?
   (b) How could you reduce the noise in the image? In doing so, how would new values for the intensity in each pixel be calculated?
   (c) Suggest one other process that could be used to enhance the image.

2 Sensing

Sensors, often miniaturised, are used to make measurements, monitor production processes and control machines. In this chapter we will give examples of:

● making microsensors
● measuring and controlling signals from sensors
● and using sensors for light, temperature, movement and strain

The examples will lead you to new ideas on electric current and potential difference, and about building electric circuits.

2.1 Making very small things

On a November day in 1960, engineer William McLellan carried a large grocery carton into a laboratory at the California Institute of Technology. The year before, in a talk entitled "There's plenty of room at the bottom", the physicist Richard Feynman had offered a $1000 prize to anyone who could make an electric motor no bigger than half a millimetre on a side. The box looked absurdly big to hold such a motor. But what McLellan took from the box was not a motor but a microscope; through the microscope Feynman watched McLellan's tiny motor turning. Feynman wrote him a cheque.

Two years earlier, in 1958, another engineer, Jack Kilby of Texas Instruments, had written in his laboratory notebook: "Extreme miniaturisation of many electrical circuits could be achieved by making resistors, capacitors and transistors and diodes on a single slice of silicon."

Kilby and others filed patents for making these miniaturised circuits – the first integrated circuits. Feynman's vision of the huge potential of making tiny circuits and tiny machines was already coming true. The era of microtechnology had begun.

Today, millions of microchips run unseen inside personal computers, calculators, domestic

A miniature electric motor. This motor, smaller than the one William McLellan showed Richard Feynman, is approximately the thickness of a human hair (0.1 mm).

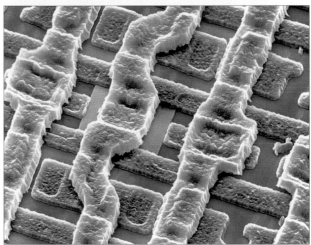

False-colour image of a computer memory chip. Pathways for electric currents crisscross the chip. The conducting strips are 3 μm wide. Today they would be a tenth of that width.

A flexible bar of silicon carrying a sharp tip. The bar is about 20 μm wide and about 5 μm thick. The tip is at most 1 μm across. The bar is part of an atomic force microscope, which scans the tip over a surface. The bar bends as the tip is attracted to the surface by intermolecular forces. The bending indicates the height of the surface below the tip. Objects on a molecular scale can be resolved.

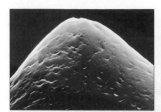

A diamond stylus finished to a sharp point. Here you see the result of mechanical polishing. The tip is sharp to about 5 μm.

Here you see a much sharper tip, about 0.1 μm across, produced by bombarding the surface with energetic ions.

Machining a tip with a beam of ions

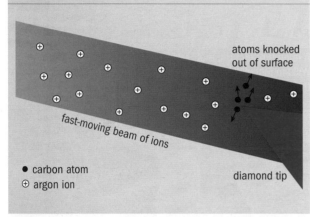

atoms knocked out of surface

fast-moving beam of ions

diamond tip

● carbon atom
⊕ argon ion

Machining with a beam of ions. Argon ions are made by firing electrons at argon atoms. The ions are then accelerated into a beam that strikes the work piece and knocks atoms out of the material. The material's shape is controlled by varying the angle of the surface to the beam of ions. Material can be removed at a rate of up to 0.1 μm per minute. The process is like using sand-blasting to clean surfaces, but on an atomic scale.

appliances and motor vehicles. Feynman was right: there is plenty of room at the bottom.

Miniature circuits and devices have completely changed the world of measurement and control. Today, a family car contains hundreds of microsensors, each one monitoring how the car is performing. One of them, no larger than a printed letter on this page, sits in the steering column checking continually for rapid deceleration, which would indicate a crash. In that event, the microsensor triggers the release of air bags to protect driver and passengers.

In a hospital intensive-care ward, microsensors that are attached to a patient monitor breathing, blood pressure, heartbeat and other vital functions. Signs of trouble quickly summon nurses and doctors to help. By being so small the sensors can be unobtrusive and convenient. Tiny microsensors help to watch over delicate premature babies

Key summary: ion and current calculations

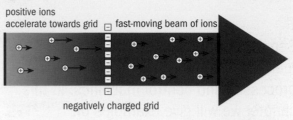

positive ions accelerate towards grid — fast-moving beam of ions

negatively charged grid

charge on ion = q electric current $I = Nq$ number N ions arrive per second

How many ions hit an atom every second in this beam?

Typical beam data
beam diameter 10 mm
(beam area = 100 mm² approximately = 10^{-4} m²)
beam current $I = 20$ mA $= 20 \times 10^{-3}$ A

Data about the ions
charge on argon ion = 1.6×10^{-19} C

1. Number of ions in beam
Number of ions arriving per second $N = \dfrac{I}{q} = \dfrac{20 \times 10^{-3}\,\text{A}}{1.6 \times 10^{-19}\,\text{C}}$

$= 12 \times 10^{16}\,\text{s}^{-1} = 10^{17}\,\text{s}^{-1}$ (approximately)

2. Area covered by one atom
Area covered by one atom $= (0.1\,\text{nm})^2 = 10^{-20}\,\text{m}^2$

3. Number of atoms in area of beam

$= \dfrac{\text{area of beam}}{\text{area of atom}} = \dfrac{10^{-4}\,\text{m}^2}{10^{-20}\,\text{m}^2} = 10^{16}$

4. Number of ions hitting an atom per second

$= \dfrac{\text{number of ions per second in beam}}{\text{number of atoms in beam}}$

$= \dfrac{10^{17}\,\text{s}^{-1}}{10^{16}} = 10$ ions per second

The very large and very small numbers come together to give a simple and remarkable result: in this beam an atom is bombarded by about 10 ions each second.

without causing them unnecessary stress. A microsensor, with communications attached, may also be implanted under a patient's skin to warn of a change in vital signs. Sensors attached to animals in the wild can detect movements of herds of deer or flocks of birds.

A way to make these tiny sensors out of silicon and other materials is to deposit material on a silicon surface and then to selectively cut away parts with acid and other chemicals. That is how

This image, showing a flash of lightning together with its spectrum, was obtained by putting a diffraction grating in front of the camera lens. The red colour is from hydrogen, the green from nitrogen and the blue from oxygen.

a miniature silicon bar used in an atomic force microscope, shown on p29, was made. Most microchips are also made in this way.

A very different way to shape miniature objects is to bombard them with a beam of fast-moving charged particles. It works very much like a sand-blaster stripping paint and rust from car bodies. Rapidly moving ions remove surface atoms, knocking them out of the surface, making it possible to smooth and shape a material right down to the near-atomic scale. Diamond tips used as super-sharp blades for surgery can be manufactured like this. The process is called ion-beam machining.

Beams of moving ions

A beam of moving charged particles carries an electric current without wires. The beam must travel in a vacuum, otherwise it will be scattered by colliding with gas molecules. Other examples of moving charged particles are:

- the beam of electrons in a dental X-ray machine;
- a beam of protons in a particle accelerator used to probe the nature of matter;
- charged particles coming from the Sun causing the Northern lights (aurora borealis).

Such a beam carries energy because the particles are moving. Moving particles do what moving particles usually do: they knock into things, delivering energy to a target. The beam carries an electric current because the particles are electrically charged and moving.

Key summary: potential difference examples

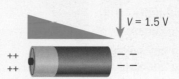

A dry cell. The chemical reaction drives electrons to one pole leaving positive charge at the other. The difference in distribution of charges creates a potential difference between the poles.

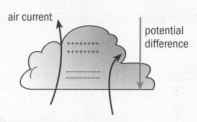

A storm cloud. Strong air currents rub ice crystals against one another, separating charges. The potential difference between the top and bottom of the cloud can be millions of volts.

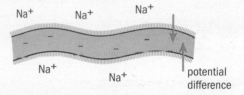

A nerve cell. "Pumps" in the walls of the cell drive sodium ions outside the cell. This difference leads to a potential difference of about 70 mV between the inside and outside. Changes in this potential difference provide the electrical signals for our nerves.

Uneven distributions of charge make potential differences

Potential difference

What makes charged particles move? It is other charged particles attracting or repelling them. The positive argon ions used for ion-beam machining are accelerated by being pulled towards a concentration of negative charge on the grid.

In the early 1800s, alone in the Nottinghamshire countryside, a young self-taught mathematician George Green was thinking about the energy of electric charges. He thought of a simple way to work out which way a charge will be pushed or pulled by concentrations of other charges. He realised that you do not need to know where all these other charges are, all you need to know is the potential energy of the charge. The charge will be pushed or pulled in the direction in which the potential energy decreases.

Calculating current, energy and power

Symbols and units

Q = charge in coulombs (C)

I = current in amperes (A)

V = potential difference in volts (V)

E = energy in joules (J)

P = power in watts (W)

t = time in seconds (s)

Δ means "change in". It always goes with another quantity, for example ΔQ is change in charge.

Basics

Electric current is the rate of flow of charge.

$$I = \frac{\Delta Q}{\Delta t}$$

Electric potential difference V between two places is the difference in potential energy ΔE of a charge ΔQ going between those places, per unit of charge.

$$V = \frac{\Delta E}{\Delta Q}$$

Power P is the rate of delivery of energy E.

$$P = \frac{\Delta E}{\Delta t}$$

Calculating formulae

Energy E given to charge Q on an ion going through a potential difference V:

$$E = QV$$

If N ions arrive per second then:
current I in beam is charge delivered per second.

$$I = NQ$$

Power P delivered by beam is energy delivered per second.

$$P = NE = NQV$$

Power delivered by a current I going through a potential difference V:

$$P = IV$$

An ion beam from an accelerator. As the ion beam exits the machine it glows blue. The beam ionises air molecules, which give out a blue light as they recombine.

This idea can be illustrated by the movement of a boulder on a hill. Going uphill, potential energy increases as the boulder gets farther from the Earth – a concentration of mass. Going downhill, potential energy decreases and the mass is pulled downhill by gravity. Green's idea was that charges go downhill too: down "electrical hills".

Green gave the name "electric potential" to the potential energy divided by the charge. Between two places there can be a difference in the electric potential: **a potential difference**. Charges move through differences in potential.

Having thought of all this by himself, Green became a student at the late age of 40. Not surprisingly his Cambridge College soon made its student a Fellow. His ideas (Green's functions) are still very important in modern theoretical physics.

Potential differences are measured in volts. One volt is one joule of energy per coulomb of charge. A thundercloud may generate a potential difference of millions of volts.

Energy and power in a beam of ions

Beams of moving ions have many uses. Ions accelerated by going round and round inside a cyclotron are used to make radioactive tracers for medical investigations. Ion beams are used to implant ions in silicon to make it conduct electricity better. An ion-beam engine firing a jet of xenon ions has been proposed for accelerating space craft making interplanetary journeys. Focused beams of ions are used to cut out shapes in tiny devices for use in micromachines.

Such tasks require energy. In a beam accelerated by a potential difference, an ion gains kinetic energy. Ion beams carry energy.

Quick check

Useful data: the magnitude of the charge on an electron, or on a singly charged ion, is 1.6×10^{-19} C

1. Show that an electron beam carrying a current of 1 µA must be switched on for 10 s to deliver a charge of 10 µC.

2. A particle accelerator delivers a million singly charged ions to a target every second. Show that the beam current is 0.16 µA.

3. The electron beam in an X-ray set is given energy by travelling across a potential difference of 10 kV. Show that the beam current to deliver $50\,\mathrm{J\,s^{-1}}$ (i.e. 50 W) to the target anode is 5 mA.

4. The energy given to an electron as it goes from one battery terminal to the other is 2.4×10^{-19} J. Show that the e.m.f. generated by the battery is 1.5 V.

5. An electron given energy by travelling across a potential difference of about 20 V can knock another electron out of an argon atom (i.e. it can ionise it). Show that the energy it has just before the collision is about 3.0×10^{-18} J.

6. An electric kettle is labelled 230 V, 2.3 kW. Show that it draws a current of 10 A.

Links to the *Advancing Physics* CD-ROM

Practise with these questions:
20S   Short answer *Revision questions*
30S   Short answer *Kinds of light bulb*
70C   Comprehension *Electron beams*
40S   Short answer *Ions in chemical cells: large and small numbers 1*

Try out these activities:
30S   Text to study *Help with calculating current and power in an ion beam*
90S   Software-based *Calculating with powers of ten and logarithms*

Look up these key terms in the A–Z:
Electric charge; electric current; electrical power; ion; potential difference

Go further for interest by looking at:
10S   Computer screen *Images of micromachined structures*

Revise using the revision checklist

2.2 Miniature circuits

"...there are now more transistors made every year than raindrops falling on California, and it costs less to produce one than to print a single character on this page." (Michael Riordan and Lillian Hoddeson 1997 *Crystal Fire*.)

The factories of the worldwide microchip industry that build tiny electric circuits in silicon are places where purity and cleanliness are paramount. Silicon crystals are grown there with impurities of less than one part in a billion. Then, layer after layer, the silicon is doped with other elements; new layers of material are evaporated onto the surface; protective layers are added and then cut away selectively; and chemicals are used to etch shapes into the material.

The tiniest particle of dust can ruin the emerging circuits. One flake of dandruff spells disaster. So the chips have to be made in ultraclean environments. The air in the room must be filtered, the doors need air-locks, and people must wear special clothing.

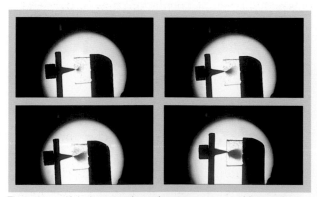

A picture of the unseen dirt around us (false-colour image). The dust mite in the middle, obliged to scramble among dust boulders, measures about 0.1 mm in length.

These beautiful photographs, taken at one second intervals, show a blue colour spreading in a heated transparent crystal of potassium chloride. There is a potential difference of about 300 V between the needle and the support of the crystal. Electrons are injected into the crystal from the needle. Trapped in spaces where an ion is missing in the crystal, the electrons absorb yellow light and the crystal looks blue. You can see the colour spreading as electrons move through the crystal.

Charges moving in conductors

A wire carrying a current does not look as if anything is going on inside, but charges are indeed on the move in it. In liquids, the drift of charges can be seen when a current flows if the current is carried by ions that colour the liquid.

In metals, electric currents are carried by moving electrons. One good reason to think that there are electrons inside metals is that electrons "boil" out of metals when the metal heated. The most common source of an electron beam, whether in a dental X-ray tube or an electron microscope, is a heated oxide-coated cathode. It gives out electrons which are then accelerated and formed into a beam by a positively charged anode.

In electroplating using a solution of copper sulphate, the current is carried by positive copper ions and by negative sulphate ions. Ions move from one electrode to the other. In red neon advertising signs, current is carried by neon ions.

The mobile charges that carry currents in silicon are mostly put there by the chip-maker. They are installed by doping the silicon with phosphorus or boron. See chapter 5 for more about how materials conduct electricity.

Rivers and electric currents

energy to water as it goes downhill

water current

height difference

energy from water as it moves material on the river bottom

energy to charge carrier as it goes down the potential hill

electric current

potential difference

energy from charge carrier as it gives energy to the material

Key summary: conductance and resistance

Conductance G
Conductance is the current for a given potential difference:

$$G = \frac{I}{V}$$

$$\text{conductance} = \frac{1}{\text{resistance}}$$

$G = 0.4\,\text{mA V}^{-1}$
$G = 4 \times 10^{-4}$ siemens

example: a bar of silicon

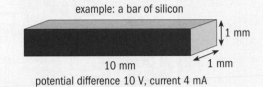

10 mm
1 mm
1 mm

potential difference 10 V, current 4 mA

Resistance R
Resistance is the potential difference needed for a given current:

$$R = \frac{V}{I}$$

$$\text{resistance} = \frac{1}{\text{conductance}}$$

$R = 2.5\,\text{V mA}^{-1}$
$R = 2.5 \times 10^{3}$ ohm

Conductance and resistance are the inverse of one another

Key summary: Ohm's law

Conductance or resistance can be calculated at any given current or potential difference. If the current or potential difference change, the conductance and resistance may change.

If the conductance and resistance are constant, independent of the current or potential difference, the conductor is said to obey Ohm's law, or to be "ohmic".

Thus, Ohm's law says that the conductance and resistance of a given component is constant. The same value can be used in calculations whatever the current or potential difference.

Most metals are ohmic at a constant temperature, however, ionised gases are not.

Only ohmic resistors have constant resistance, independent of current

Getting charge carriers moving

Like molecules in a gas, the mobile charge carriers in a conductor move about continually in all directions. For a current to flow, they have to be made to move together in one direction on average. It is a potential difference that gets this drift going in one direction.

An electric current is rather like water flowing in a river. The water in a river does not necessarily speed up as it goes downhill. Energy gained by the water from going downhill is given up again as the moving water forms eddies and rubs against the river bed. Energy leaves the moving water as fast as it enters.

The same happens with electric currents. In a conductor the drift of mobile charge carriers is steady. They gain energy by falling down a potential difference and give up energy again to the conducting material in which they are moving. Energy flows in from the potential difference and

out again to heat up the conductor.

The charge carriers in a conductor do not all start moving the instant the potential difference is switched on. It is easy to show that a signal takes about 1 μs to travel the length of a 300 m reel of cable. When the potential difference is switched on, an electromagnetic impulse sweeps round the circuit, leaving behind potential differences along the wires that keep the charge carriers moving. Computer designers have to worry about these delays: hundreds of calculations may be over and done with in 1 μs. One reason to build microchips small is that it reduces such delay times.

How much current from a potential difference?

The designer of a microchip needs a quantity that says how much current flows for a given potential difference. The amount differs from one conductor to another. Some conductors with many charge carriers that move easily give a large current for a small potential difference. Others, with few charge carriers able to move, only carry a small current. The conductance G measures how well a wire, a chip component or an electroplating cell conducts:

$$\text{conductance } G = \frac{\text{current } I}{\text{potential difference } V}$$

Key summary: power equation

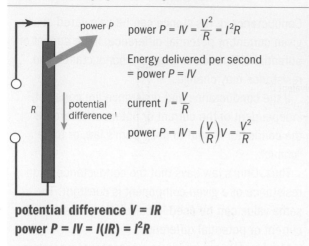

$$\text{power } P = IV = \frac{V^2}{R} = I^2 R$$

Energy delivered per second
$= \text{power } P = IV$

$$\text{current } I = \frac{V}{R}$$

$$\text{power } P = IV = \left(\frac{V}{R}\right)V = \frac{V^2}{R}$$

potential difference $V = IR$
power $P = IV = I(IR) = I^2 R$

Key summary: series and parallel rivers

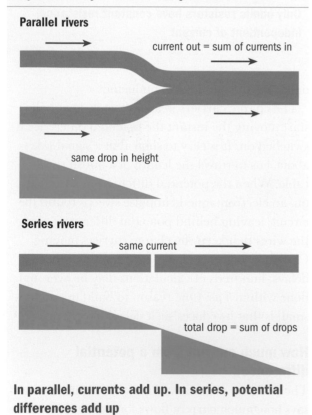

Parallel rivers

current out = sum of currents in

same drop in height

Series rivers

same current

total drop = sum of drops

In parallel, currents add up. In series, potential differences add up

The unit of conductance is amperes per volt, or siemens (S).

The conductance of an object is not necessarily the same whatever the current or potential difference. The conductance would increase if the moving charges started to hit and ionise other atoms in the material, thus giving more mobile carriers.

A microchip is a fantastically complicated set of electric circuits all built on a piece of silicon. This one is small and simple compared with a more modern computer chip, and so you can look at the whole chip and still see some of the connections.

Circuit designers also think in terms of how badly things conduct.

$$\text{resistance } R = \frac{\text{potential difference } V}{\text{current } I}$$

$$\text{resistance } R = \frac{1}{\text{conductance } G}$$

Resistance is given in volts per ampere, or ohms (Ω).

If you know the conductance, you can immediately work out the resistance. Told the resistance, you can immediately work out the conductance. It's like switching from thinking about the speed of the car to thinking about how long the trip will take. Pairs of reciprocal quantities like these are common in physics.

Microchips run hot

Microchips in computers often run hot, and computers need cooling fans to stop chips overheating. The reason is that circuit components are packed very densely. Each dissipates little power, but the number of components in a given area is very large. The result: some supercomputers have to be cooled with liquid nitrogen.

For other purposes – heating and lighting – the power dissipated when a potential difference makes a current flow is just what is wanted.

Key summary: conductors in parallel and series

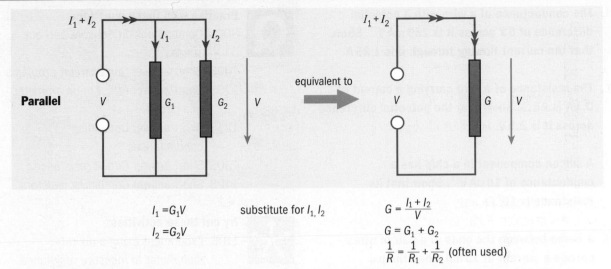

Parallel

$I_1 = G_1 V$

$I_2 = G_2 V$

substitute for I_1, I_2

$G = \dfrac{I_1 + I_2}{V}$

$G = G_1 + G_2$

$\dfrac{1}{R} = \dfrac{1}{R_1} + \dfrac{1}{R_2}$ (often used)

Currents add up, potential difference is the same for both: conductances add up. Example: lamps in domestic wiring

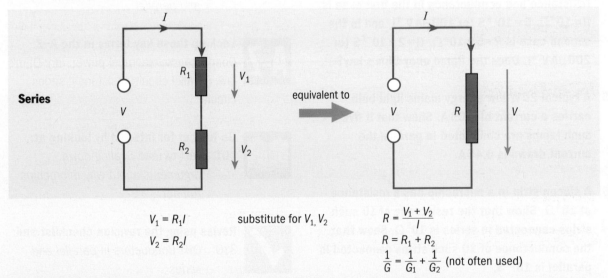

Series

$V_1 = R_1 I$

$V_2 = R_2 I$

substitute for V_1, V_2

$R = \dfrac{V_1 + V_2}{I}$

$R = R_1 + R_2$

$\dfrac{1}{G} = \dfrac{1}{G_1} + \dfrac{1}{G_2}$ (not often used)

Potential differences add up, current is the same for both: resistances add up. Example: potential divider (see section 2.3).

Connecting conductors together

Connections on a silicon chip can be complicated. The same source of potential difference may have to send current through several components at once. The two main kinds of connection are parallel and series.

In parallel, components provide alternative side-by-side conducting paths. Charges coming in to a junction all have to go out again, so the currents in each parallel branch add up to the current coming in to them all. The potential difference across the

components is the same. More water can go down two rivers side by side on the same hill than can go down one. Thus for components in parallel, the conductances add up.

In series, components are connected one after the other, and each component requires a part of the potential difference to drive charge through it. The current is the same in both but has farther to go. A greater drop is needed for a longer river to carry the same flow of water. Thus for components in series the resistances add up.

Quick check

1. The conductance of a wire with a potential difference of 5 V across it is $250 \, \mathrm{mAV^{-1}}$. Show that the current flowing through it is 1.25 A.

2. The resistance of a wire carrying a current of 0.1 A is $25 \, \Omega$. Show that the potential difference across it is 2.5 V.

3. A silicon component in a chip has a conductance of $10 \, \mathrm{\mu AV^{-1}}$. Show that its resistance is $10^5 \, \Omega$.

4. A flame between the ends of a pair of wires carries a current of 10 mA when the p.d. across the wires is 100 V, and a current of 40 mA when the p.d. is 200 V. Show that the resistance and conductance in the first case is $R = 10^4 \, \Omega$, $G = 10^{-4} \, \mathrm{S}$ (or $100 \, \mathrm{\mu AV^{-1}}$), and in the second case is $R = 5 \times 10^3 \, \Omega$, $G = 2 \times 10^{-4} \, \mathrm{S}$ (or $200 \, \mathrm{\mu AV^{-1}}$). Does the flame obey Ohm's law?

5. A typical 20 W low-energy mains light bulb carries a current of 0.09 A. Show that if five such lamps are connected in parallel the current drawn is 0.45 A.

6. A silicon strip in a microchip has a resistance of $10^4 \, \Omega$. Show that the resistance of 10 such strips connected in series is $10^5 \, \Omega$. Show that the conductance of 10 such strips connected in parallel is $10^{-3} \, \mathrm{S}$.

Links to the *Advancing Physics* CD-ROM

Practise with these questions:

80C   Comprehension *Sensors and our senses*

100S   Short answer *Some circuit problems*

120S   Short answer *Effect of an ammeter in a circuit*

125S   Short answer *Combining conductances*

130S   Short answer *Circuit resistance*

140S   Short answer *Combining resistors*

Try out these activities:

110E   Experiment *Using a digital multimeter to measure resistance: Plot and look*

120E   Experiment *Resistors in series and parallel*

Look up these key terms in the A–Z:

Conductance; electrical power; ion; Ohm's law; parallel circuit; resistance; series circuit

Go further for interest by looking at:

40T   Text to read *Scaling-down arguments about why microchips run hot*

Revise using the revision checklist and:

310   OHT *Conductors in parallel and series*

320   OHT *Series and parallel rivers*

330   OHT *Rivers and electric currents*

340   OHT *Potential difference: examples*

350   OHT *Calculating conductance and resistance*

2.3 Controlling and measuring potential differences

Radio and hi-fi

To turn up the volume on a radio, you turn a knob, its volume control. On an old radio there may be a scratchy sound as you turn the knob. What is going on behind? The signal from the radio is controlled by tapping off part of it with a sliding contact moving along the surface of a high resistance. The high resistance may be a film of carbon or a wire coil. The device is called a potential divider, or sometimes a "potentiometer" and even "pot" for short. It works by having the signal from the radio – a varying potential difference – across the whole resistance, but the signal to the radio loudspeaker taken from across only part of the whole resistance. The scratchy noise on an old radio may come from dirt on the surface of the resistance, which briefly spoils the contact of the slider as it moves. Top-quality high-fidelity equipment sometimes uses a chain of fixed resistors between gold-plated contacts to beat the noise problem.

In the top-quality volume control, all of the

Key summary: potential divider used to control a potential difference

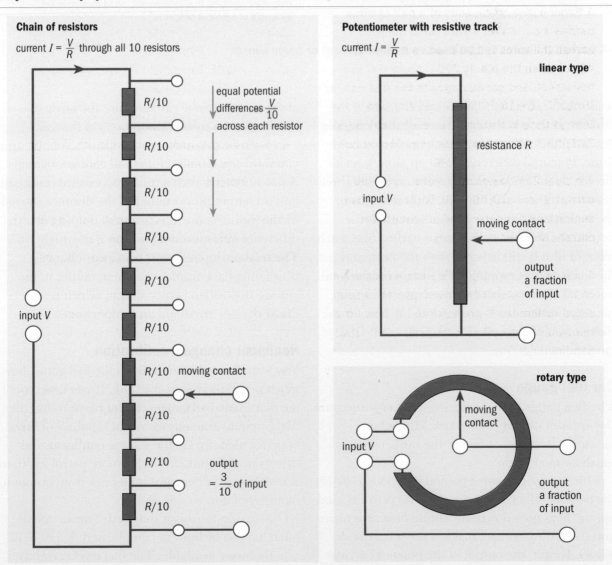

Chain of resistors

current $I = \dfrac{V}{R}$ through all 10 resistors

equal potential differences $\dfrac{V}{10}$ across each resistor

$R/10$ (×10)

input V

moving contact

output $= \dfrac{3}{10}$ of input

Potentiometer with resistive track

current $I = \dfrac{V}{R}$

linear type

resistance R

input V

moving contact

output a fraction of input

rotary type

moving contact

input V

output a fraction of input

A potential divider taps off a fraction of its input to provide a controlled output. Equal movements of the sliding contact give equal changes in output.

Key summary: potential divider as a displacement sensor

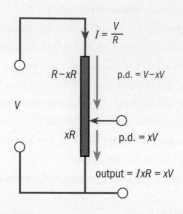

sliding contact displaced by fraction x of
length of potential divider

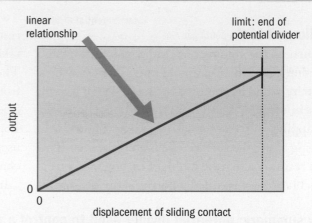

output proportional to displacement, up to maximum set
by length of potential divider

A potential divider can be used as a displacement or angle sensor

resistors are in series. They carry the same current. If their resistances were all the same, the potential difference across each one would be the same. Potential differences add up along a series circuit, so the potential difference across say three resistors, if there are 10 in all, is three-tenths of the potential difference across them all. If the potential divider is a continuous carbon film, each piece of film is still in series with the next, and the same idea of tapping off a fraction of the total potential difference still works. Now the output potential difference is proportional to how far the slider has been moved along – or around – the potentiometer.

Car fuel gauge

The fuel gauge of a car does not actually measure the amount of fuel in the tank. Instead, it measures the level of fuel in the tank, which is much simpler to do.

A fuel gauge can use a potential divider to detect the position of a float in the tank. As petrol is used up, the float goes down, the tilting float arm turns and the sliding contact inside a potential divider moves. Result: the output of the potential divider changes and a signal goes to the fuel gauge. All too soon the driver is looking for a petrol station. In a similar way, a position sensor can keep track of how far a robot's finger is from what it is touching,

and an angle sensor can measure the angle of bending of the robot's elbow.

In a sensor measuring the angle of a robotic arm you want equal output for equal changes in angle. A potentiometer with a uniform resistive track can give an output proportional to the distance moved by the sliding contact. A graph of output potential difference against slider position is a straight line. The relationship is linear. For linear changes, translating back from a change in output to the change in position is very simple, which is why linear devices are useful and important.

Nonlinear changes: calibration

You want to read straight from the fuel gauge how much petrol there is in the tank. If you look, you'll see that many fuel gauges do not move round the dial by equal amounts for equal numbers of litres of petrol used up, i.e. the scale is nonlinear and drivers might think they have more petrol left than is really there. The translation back from output to input is not now so simple.

One reason for this is that petrol tanks in cars often have to be built in complicated shapes to fit into the space available. The fuel level certainly goes down as petrol is used up, but changes in the level have no simple relationship to the amount of fuel used. Thus the relationship between float level and amount of fuel is nonlinear. A graph

A car fuel gauge

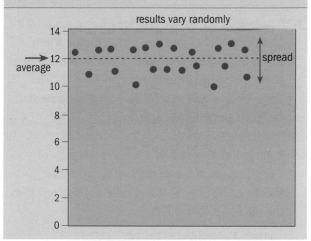

A float connected to an arm rises and falls with the fuel level, and the arm turns, moving a contact round a resistor in the form of a circle. The resistor is a potential divider acting as a sensor.

of this relationship would have a strange shape – certainly not a straight line. Such a graph is called a calibration curve. You will need one of these for any sensor that you make or buy. A calibration curve tells you how to look up the input to a measurement system if you know its output. Calibration curves are now often built into sensor systems as look-up tables relating input to output.

Look-up tables are especially useful when a sensor is nonlinear. This is part of a general trend: software replacing hardware. Mass producing large numbers of carefully designed hardware items is expensive; copying software once it is written is extremely cheap.

Results can fluctuate

Noise, such as that from a dirty volume control, is often present in measurements. For example, the level of fuel in a petrol tank will vary as the fuel sloshes around. A sluggish gauge might not notice; but one with a fast response time would need to take frequent sample readings and calculate an average over time.

This is an example of a general problem in instrumentation: results can vary around the true value in an unsystematic way. Here are some other examples of this:

- radioactive carbon dating in archaeology: the number of decays recorded in a given time varies randomly;
- noise in astronomical images (chapter 1);
- fluctuations in worldwide temperatures making it hard to see that the greenhouse effect is actually raising the average temperature.

Noise in images can be reduced by averaging

Variations in results

Results can vary about the true value. If the variations are random, their average will be a better indication of the true value than any one result alone. But errors need not be random, they can be systematic (i.e. results are consistently wrong in one direction) or they can be accidental. The average of a set of results varies less than individual results vary.

values of neighbouring pixels (chapter 1) – this is fundamentally the same idea as averaging readings that fluctuate. Taking the average works well if fluctuations are unsystematic. The average varies less than any one result.

Small variations are usually most important for sensitive instruments. When physicist R V Jones, who led much of the scientific effort during the Second World War, built an extremely sensitive tilt detector at the University of Aberdeen, which is on the coast, he found that he was able to detect the university building tilting as the tide came in.

These examples show that errors or variations in readings are sometimes random in nature – as with radioactive decays – and sometimes systematic, for example when the tide comes in and the ground tilts under its weight. The sloshing of petrol in a tank is probably not really random (it is quite like the tide going in and out) but it is complicated and unsystematic.

In your own experiments there are many such sources of variation. Watching a digital meter you often see the last digit varying. Repeated readings rarely agree exactly. The result of a measurement is only complete when, besides reporting the average of a set of results, you also report its uncertainty – that is, the range over which you think the result could vary.

Quick check

1. A potential divider 50 mm long has a p.d. of 2 V across it. The sliding contact is moved 5 mm along the potential divider. Show that the output of the potential divider changes by 0.2 V. The slider is now moved a further 10 mm. Show that the change in output is 0.4 V.

2. A potential divider has a total resistance of 1 kΩ, and carries a current of 3 mA. If the sliding contact is at the middle point, show that the output p.d. is 1.5 V.

3. The swinging of a pendulum is detected by a circular rotary potential divider. The output displayed on an oscilloscope has a variation from peak to trough of 100 mV. The input to the potential divider is 1 V. Show that the angle through which the pendulum is swinging is about 36 degrees.

4. In use, a potential divider becomes hot and its resistance increases from 10 to 11 kΩ. If the input p.d. is 2 V, show that the current through it changes by 0.02 mA. Explain why, if the sliding contact stays in the same place, the output p.d. does not change.

5. Sketch a V-shaped petrol tank. Use your sketch to construct a look-up table to convert levels in the tank to volumes in the tank.

6. If you measure the resistances of each of a large batch of supposedly identical commercial resistors, you get a spread of results. If you measure the intensity of pixels in an area of an image that is supposedly uniformly bright, you get a spread of results (chapter 1). Explain the connection between averaging a set of experimental results and smoothing an image.

Links to the *Advancing Physics* CD-ROM

Practise with these questions:
170S Short answer *Tapping off potential difference*
180S Short answer *Loading the potential divider*
190D Data handling *Lamp and resistor in series*
200S Short answer *Controlling a robot arm*

Try out these activities:
200E Experiment *Potential dividers*
210S Software-based *Effect of load on a potential divider*

Look up these key terms in the A–Z:
Accuracy; average; calibration; linear device; potential divider; systematic error; uncertainty

Go further for interest by looking at:
70T Text to read *When is a potential divider linear?*

Revise using the revision checklist and:
1450 OHT *Sources and internal resistance*
60S Text to study *The potential divider in pictures*

2.4 Sensors and our senses

Would you like to live in a "smart" home that recognises you when you come in and puts on the kettle for you? Perhaps intelligent surveillance systems are more likely to get here first. What sensing systems would they need? Sight? Hearing? Touch? Smell? Taste? All of these are possible electronically.

Cars are now largely built by robots. A few people are still needed, but mainly to look after the robots. The robots must be able to sense, to fractions of a millimetre, how accurately they are putting parts of the car in place. The cars they build contain hundreds of sensors, for example to detect and prevent skidding, among other things. In a car what do sensors need to measure or detect? Position? Speed? Acceleration?

Information about the world can be obtained in many ways. Your senses use only a few of them, but have the advantage of being made smart by being connected to a powerful brain. Modern sensors pick up information that a human cannot sense, but they have only recently started to get even a little smart. An ultrasonic distance sensor is one example. It can measure temperature, which affects the speed of sound waves, and then compensate for it.

Electronic ears

A microphone is an electronic ear, producing a varying electrical signal from pressure changes caused by sounds. On p3 you can see a piezoelectric crystal, such as quartz, being used to detect high-frequency (ultrasonic) sound. With a piezoelectric gas lighter, you squeeze it and get a spark. Both devices generate their own potential difference. Piezoelectric crystals usually have a high resistance. Almost no current flows in a circuit connected to a high-resistance source like this.

Electronic skin

Your skin can detect changes in temperature and pressure. One way to detect temperature changes electrically is to use a thermocouple. A thermocouple is just a pair of different metals joined together: when the junction is made hot a potential difference is created (the Seebeck effect). The potential difference is small (a

This robot was able to sense, to fractions of a millimetre, how far to move its hand to get the slinky running.

This radiation sensor is a tiny 3×3 mm chip. Infrared radiation falls on the chip and is absorbed by a thin 0.55 mm-diameter black disc at the centre of the chip, thus warming the disc. Grids of thermocouples measure the temperature of the disc.

few millivolts or less), but the resistance of a thermocouple is generally small, so despite the small potential difference it will give a measurable current.

Different pairs of metals produce different thermoelectric e.m.f.s. One of the largest is $120\,\mu V\,°C^{-1}$ obtained from the pairing of antimony and bismuth. Using multiple junctions with alternating hot and cool junctions increases the effect. Ten antimony–bismuth junctions in series give 10 times the e.m.f. You can see four grids of alternating bars of antimony and bismuth in the photograph of the radiation sensor. In this way, the sensitivity of the detector can be increased. The **sensitivity** of a sensor is the ratio of change in output to change in input, in this case a few millivolts per degree.

Getting a signal out

The examples of a crystal microphone to detect sound and of a thermocouple to detect warmth show that there can be problems in detecting a signal from a sensor. The thermocouple needs a sensitive voltmeter, able to detect a signal of as

Key summary: internal resistance and e.m.f.

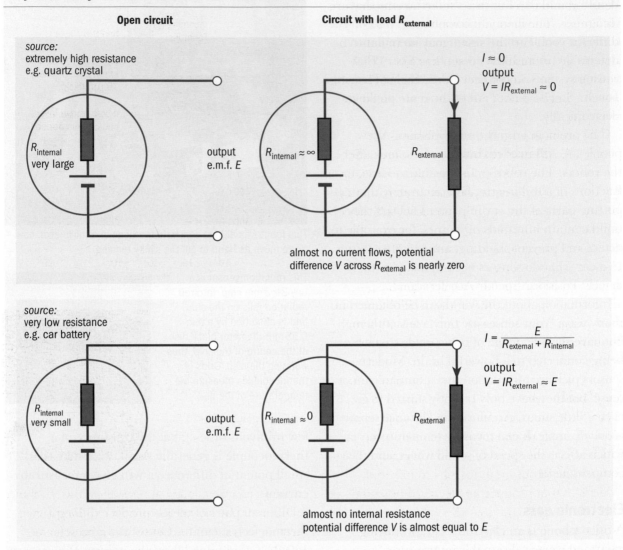

Open circuit

Circuit with load $R_{external}$

source:
extremely high resistance
e.g. quartz crystal

$R_{internal}$ very large

output e.m.f. E

$R_{internal} \approx \infty$

$R_{external}$

$I \approx 0$
output
$V = IR_{external} \approx 0$

almost no current flows, potential difference V across $R_{external}$ is nearly zero

source:
very low resistance
e.g. car battery

$R_{internal}$ very small

output e.m.f. E

$R_{internal} \approx 0$

$R_{external}$

$I = \dfrac{E}{R_{external} + R_{internal}}$

output
$V = IR_{external} \approx E$

almost no internal resistance potential difference V is almost equal to E

source:
some internal resistance
e.g. dry cell, photocell

$R_{internal}$

output e.m.f. E

$R_{internal}$

$R_{external}$

$I = \dfrac{E}{R_{external} + R_{internal}}$

output
$V = IR_{external} = E - IR_{internal}$

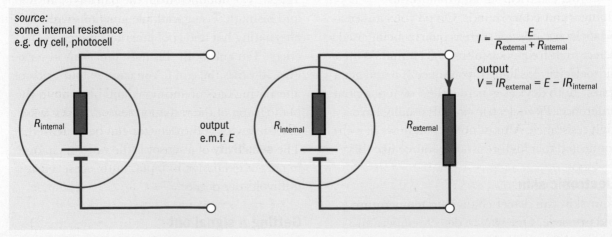

Potential differences add up: $E = V_{external} + V_{internal}$

$\qquad E = IR_{external} + IR_{internal}$

$\qquad V_{external} = IR_{external} = E - IR_{internal}$

The electromotive force (e.m.f.) is equal to the potential difference from a source on open circuit

little as one-thousandth of a volt. Too insensitive a detector, and there appears to be no temperature difference. The crystal microphone gives a different problem: the signal may seem not to be detectable because the crystal has a very high resistance. So you connect a detector, but see nothing because not enough current can flow to drive the detector.

The problem with the microphone is the problem of **internal resistance** of a source. If the source is connected to a low resistance detector or other load, the potential difference across it may be tiny, because most of the potential difference available is used in driving a small current through the source's large internal resistance. Only if the detector has a very high resistance does the output of the sensor show an appreciable potential difference. Luckily cathode-ray oscilloscopes and audio amplifiers can easily be made with a high input resistance, so crystal microphones work well with them.

Electronic eyes

The charge-coupled device (CCD) on p6 is one kind of electronic eye. Another is a silicon p–n junction, which can detect radiation over a range of wavelengths, from ultraviolet to infrared.

Devices that produce an electrical potential difference from light are called photovoltaic cells. Notice that in the graph (top, right) showing potential difference against illuminance (amount of light), illuminance is plotted on a logarithmic scale. The potential difference increases in equal steps as the amount of light falling on the photovoltaic cell is multiplied by a constant factor. This more or less matches what the human eye does.

Space craft and the Hubble Space Telescope orbiting the Earth get their power from similar photovoltaic cells, but on a grander scale. The cells are placed on wings that fold out to catch the maximum of amount of sunlight.

Photoconductive sensors

Materials such as cadmium sulphide (CdS) conduct better when light shines on them, so they make sensitive and cheap electronic eyes. They need a power supply and work by varying the current drawn from it as their conductance changes.

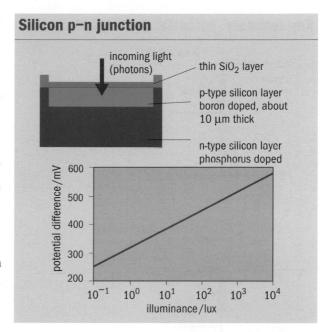

Silicon p–n junction

incoming light (photons)

thin SiO_2 layer

p-type silicon layer boron doped, about 10 μm thick

n-type silicon layer phosphorus doped

The Hubble telescope, attached to the Space Shuttle, after being repaired and refurbished. The gold coloured wing-like panels are the telescope's power supply. The photovoltaic cells on the wings generate electricity from sunlight.

To get a large change in resistance, the resistance is increased by laying down CdS in a long zigzag strip. The change in resistance can be detected using a potential divider. The output of the potential divider can be arranged to increase, or to decrease, when the sensor is illuminated.

Response time

Photoconductive sensors, although sensitive and cheap, are slow to give a result. They take about 10 ms to respond to a change in light level; a p–n diode can give a result in about 1 μs. A photoconductive sensor would not detect a rapid change in brightness, for example the change produced by a camera flash gun.

The response time of a thermocouple may be quite long, since the metal junction takes time to warm up. The response time of a clinical thermometer is the length of time you have to hold it under your tongue. Clinical mercury thermometers are now being replaced by infrared-

Key summary: potential divider/CdS sensor

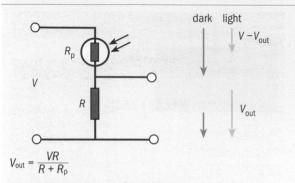

$$V_{out} = \frac{VR}{R + R_p}$$

output p.d. V_{out} increases when R_p decreases

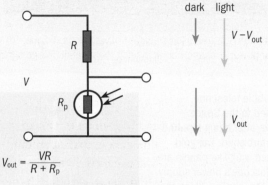

$$V_{out} = \frac{VR}{R + R_p}$$

output p.d. V_{out} decreases when R_p decreases

V = input potential difference
R_p = resistance of photocell
R = other resistance in potential divider

The output of a sensor in a potential divider can be taken across either component

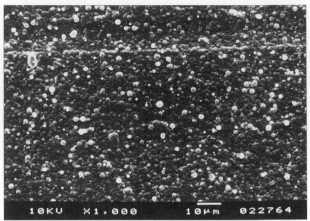

A layer of polymer polypyrrole, deposited by electrolysis, lies across a narrow gap in a microsensor. The conductance of the polymer increases when it is exposed to certain odour molecules, with a sensitivity down to 0.1 ppm.

Katherine Blodgett was the first woman to obtain a PhD degree in physics at the University of Cambridge, in 1926. As well as pioneering techniques for making what are now called Langmuir–Blodgett films, Blodgett played an important role in developing antireflection coatings for lenses, which are now standard in cameras and binoculars.

detecting ear thermometers, which have a much faster response time.

Electronic noses

The human nose can detect very small amounts of some substances. Insects can detect extremely low levels of chemical signals (pheromones) from their potential mates.

Sensors in domestic fire alarms detect smoke. Other devices can detect pollutants, tell fresh food from bad, and some can now detect diseases by "smelling" the breath of a patient. Many of these sensors detect a change in resistance caused by a gas in contact with the sensor material. Again, a potential divider can be used to provide an electrical output.

Recently, films of electrically conducting polymers have been used in microsensors. The polymer can be tailored to be sensitive to particular molecules, which is just what is needed in a good nose.

Very thin layers of other kinds of substances are also used in sensors. Very thin layers can be produced by floating a film of a material on water. This way of making layers of only a few molecules thick was pioneered in the 1930s by Katherine Blodgett and Irving Langmuir.

Gas sensors can also monitor car exhausts to check on engine performance. Sensors that detect a change in resistance are used here too.

Thermal sensing

The fact that many sensors are affected by changes in temperature means that there are many different ways to sense temperature changes.

Gas sensing using tin oxide

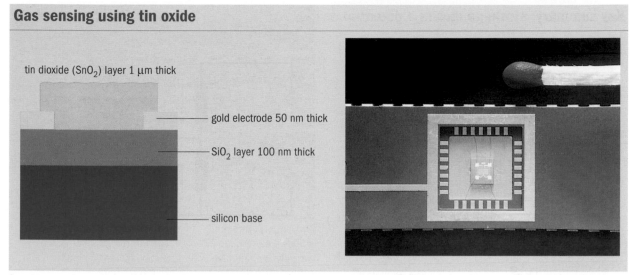

tin dioxide (SnO$_2$) layer 1 μm thick

gold electrode 50 nm thick

SiO$_2$ layer 100 nm thick

silicon base

A gas sensor for a car exhaust has to work in a hostile environment. A common material used to detect oxygen is tin dioxide. Tin dioxide detectors have to run hot (400 °C), so there is a big advantage for power consumption in making them small. Image right: a thin-fim tin dioxide gas microsensor is shown alongside a match head.

Physicists at the National Physical Laboratory often say "every sensor is a thermometer". By this they mean that whatever sensor you use, you have to consider the disturbing effects of temperature on the measurements it makes.

Thermistors are beads of metal oxides whose conductance increases rapidly with temperature. They make cheap and sensitive temperature sensors when used with a potential divider in the same way as a photoconductive sensor (p45). Their conductance does not increase linearly with temperature, so it is helpful to use a look-up table (p41) to convert the p.d. from the potential divider to give a temperature reading.

A temperature sensor also helps people with diabetes to monitor their blood-sugar levels. The glucose in a tiny drop of blood is oxidised by an enzyme on an organic film; the temperature rise shows the amount of glucose.

Strain gauges

Especially in mining areas, houses are liable to subsidence. One way to check is to fit strain gauges to structural parts of the house.

The resistance of a metal increases when it is stretched, simply because it becomes longer and thinner. This gives a way to make a strain gauge.

A strain gauge is a zigzag strip of metal foil glued to the surface of a component whose strain is to be monitored. The foil is stretched if the

component is stretched. Stretching along the length of the strips in the foil is detected as an increase in resistance. The zigzag arrangement simply increases the resistance of the gauge – usually about 100 Ω – which increases the change in resistance. Such strain gauges are cheap and convenient to use.

Suppose the bending of a beam has to be measured. Such measurements are needed in investigating the strengths of materials (chapter 4). A really clever trick is to use two strain gauges, so that as the beam bends, one is stretched and the other is compressed. Connected in a potential divider, the two gauges give twice the output that one would give alone.

With this set-up there is a bonus: the combination of the gauges is not so sensitive to changes in temperature. The resistance of a strain gauge increases with temperature. By using just one strain gauge in a potential divider, a rise in temperature cannot be told apart from a strain. Using two gauges, if the temperature goes up the resistances of both gauges increase, so the output of the potential divider does not change.

This removal of the effect of temperature is an example of something that is important in measurement. Either the apparatus is designed to eliminate the systematic effect of a disturbing influence, or the effect is measured independently and then allowed for.

Key summary: strain gauges in a potential divider

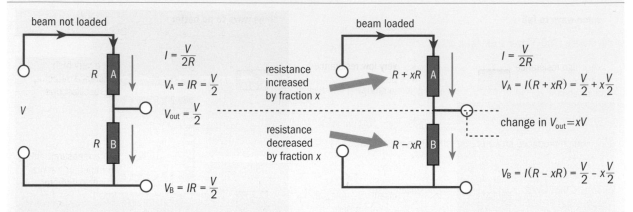

beam not loaded

$I = \dfrac{V}{2R}$

$V_A = IR = \dfrac{V}{2}$

$V_{out} = \dfrac{V}{2}$

$V_B = IR = \dfrac{V}{2}$

beam loaded

resistance increased by fraction x

resistance decreased by fraction x

$I = \dfrac{V}{2R}$

$V_A = I(R + xR) = \dfrac{V}{2} + x\dfrac{V}{2}$

change in $V_{out} = xV$

$V_B = I(R - xR) = \dfrac{V}{2} - x\dfrac{V}{2}$

Approximations
The gauges A and B have equal resistance R. The changes in resistance xR of A and B are equal and opposite.

How it works
When the beam bends, A increases and B decreases in resistance. The total resistance $2R$ does not change, so the current I does not change. A has a larger p.d. across it; B has a smaller one.

Result
The output V_{out} changes by xV.

Strain gauges can be paired to compensate for temperature changes

Strain-gauge design

These gauges are made to detect strains in several directions.

Bending a beam

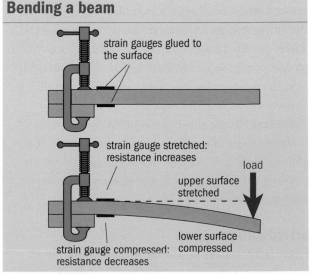

strain gauges glued to the surface

strain gauge stretched: resistance increases

load

upper surface stretched

lower surface stretched

strain gauge compressed: resistance decreases

lower surface compressed

Bending stretches the upper surface of the beam and squashes the lower surface. Strain gauges attached to these surfaces measure the amount of stretching and squashing.

Getting output from sensors

If you are building a sensor system, you have to think how to get a big enough signal out that can be recorded by the instruments that you have.

Key summary: getting signals from sensors

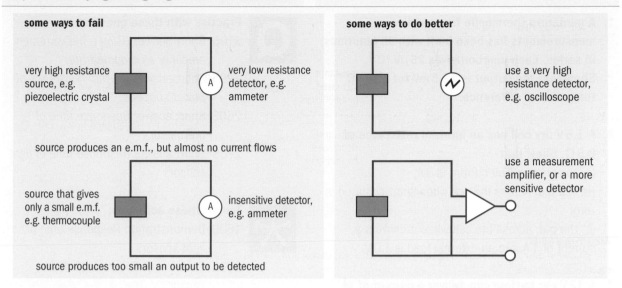

some ways to fail

very high resistance source, e.g. piezoelectric crystal — very low resistance detector, e.g. ammeter (A)

source produces an e.m.f., but almost no current flows

source that gives only a small e.m.f. e.g. thermocouple — insensitive detector, e.g. ammeter (A)

source produces too small an output to be detected

some ways to do better

use a very high resistance detector, e.g. oscilloscope

use a measurement amplifier, or a more sensitive detector

Source and detector must be matched to get an output from a sensor

There are two problems to think about.

First, how big is the signal and how sensitive is the detector? Are you trying to detect a signal of a few millivolts with an instrument that cannot see so small a quantity? Can the signal be amplified?

Second, what is the effect of the resistances in the detection circuit? Does the sensor have a large internal resistance? If so, you will get very little current from it. That will not matter if your detector needs very little current – i.e., if it too has a high resistance.

Internal resistance of source

Many sensors, being made of semiconductors, have a high internal resistance. Some, like quartz crystals, have very high internal resistance indeed, being practically insulators. When a detector tries to draw current from such a source, the current is limited by the internal resistance of the source. The potential difference across the detector becomes much less than the e.m.f. of the source.

Think about potential differences in a series circuit. The load resistance R_{load} is in series with the internal resistance $R_{internal}$ of the source. The potential difference across the two must add up to the total e.m.f. E of the source. This is just the conversion of energy: the energy per coulomb dissipated in the two resistances is equal to the

Key summary: $V = E - IR_{internal}$

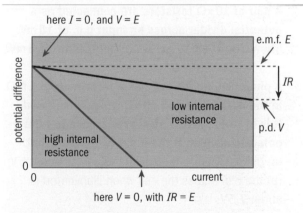

here $I = 0$, and $V = E$

potential difference

e.m.f. E

IR

low internal resistance

p.d. V

high internal resistance

current

here $V = 0$, with $IR = E$

Internal resistance reduces output potential difference when current is drawn

total energy per coulomb available.

The two potential differences are given by IR_{load} and $IR_{internal}$, so $IR_{load} + IR_{internal} = E$. Writing V for the potential difference IR_{load} across the load gives $V = E - IR_{internal}$. What does this equation say? If the internal resistance is very small. $V = E$ approximately. The largest current that can be drawn is when $E = I/R_{internal}$ at which point V has fallen to zero. Generally, the larger the current drawn, the smaller the output potential difference V.

Quick check

1. A miniature thermopile for sensitive measurements has been built with 90 junctions in series. Each junction gives 25 µV °C⁻¹. Show that the output is 22.5 mV for a 10 °C temperature difference.

2. A 1.5 V dry cell has an internal resistance of 0.5 Ω. Show that
(a) the short-circuit current is 3 A;
(b) the p.d. across the cell when short-circuited is zero;
(c) the p.d. across the cell when it delivers a current of 1 A into an external load is 1.0 V.

3. A 12 V car battery can deliver a current of at least 48 A to the starter motor. Show that the largest value that its internal resistance can have is 0.25 Ω.

4. A pair of 10 kΩ resistors are connected as a potential divider across an input p.d. of 10 mV. Show that the p.d. across either resistor is 5 mV.

5. A photoconductive cell has a resistance of 1 MΩ in the dark, falling to 10 kΩ when illuminated. It is in series with a 10 kΩ resistance to form a potential divider. If the input is 5 V show that
(a) the p.d. across the cell in the dark is about 5 V;
(b) the p.d. across the cell when illuminated is about 2.5 V.

6. Two 100 Ω strain gauges on a beam are in series as a potential divider. The input p.d. is 1 V. Show that the current is 5 mA. When the beam bends, one gauge increases in resistance by 0.1 Ω and the other decreases by 0.1 Ω. Show that the p.d. across each strain gauge changes by +0.5 mV and −0.5 mV respectively.

Links to the *Advancing Physics* CD-ROM

Practise with these questions:
210S Short answer *Using a measurement amplifier as a comparator*
220S Short answer *Internal resistance of power supplies*
260S Short answer *Response time of thermistors*
270S Short answer *Response time of light sensors*

Try out these activities:
160D Demonstration *Response time of light sensors*
220E Experiment *Calibrating a light sensor*

Look up these key terms in the A–Z:
Electromotive force; internal resistance; photocell; potential divider; response time; thermistor; thermocouple

Go further for interest by looking at:
20T Text to read *An electronic toaster which senses when the toast is done*

Revise using the revision checklist

2.5 Measuring well and knowing how well you have done

The main qualities of a good sensor system are:

- high resolution;
- appropriate output for a given input (sensitivity);
- rapid response time;
- small variations between repeated results (uncertainty);
- and small systematic or zero errors.

A sensor needs to be calibrated against a known standard. Linearity is also useful, but is today less important with the use of look-up tables.

Resolution

The resolution of a sensor is the smallest change that the sensor can detect in the quantity it is measuring. In a digital display, the least significant digit will often fluctuate, indicating that changes of that magnitude are only just resolved. Checking the resolution of an instrument is a way to estimate the uncertainty in the measurements that it delivers.

Sensitivity

The sensitivity of a measuring system is the ratio of change of output to change of input, e.g. an oscilloscope display might be stated in millimetres per volt. Sensitivity can often be increased by amplifying the output signal.

Response time

The response time is the length of time a sensor takes to reach its final reading following a sharp change in input. Changes that vary more rapidly than this will generally be averaged out. The response time always needs to be short enough to detect important changes as they occur.

Uncertainty

Uncertainty is the extent to which you can't be sure of a measurement. Any result can only be known to within a finite range of values. This is because small unsystematic or random variations (for example, noise) are present in all experimental data. The size of these variations limits the precision with which a measurement can be made. The magnitude of the uncertainty can be estimated in two ways:

- from the properties of the measurement devices, especially their resolution;

Key summary: systematic error, uncertainty

Think of measurements as shots on a target. Imagine the "true value" is at the centre of the target

small uncertainty small systematic error
precise, accurate

head this way to do better

large uncertainty small systematic error
imprecise, accurate

small uncertainty large systematic error
precise, inaccurate

large uncertainty large systematic error
imprecise, inaccurate

To improve a measurement you need to reduce both systematic error and uncertainty

- and from the results of repeated measurements, seeing how much they vary.

A simple measure of uncertainty is the range of values within which the result is expected to fall.

If measurements can be repeated under the same conditions, their average is a better indication of the true value than any one measurement on its own. The uncertainty of the average of many repeated readings is smaller than the range of the readings themselves.

When the uncertainty is small, with results scattered over a very narrow range, the precision of the experiment is said to be high, but this does not mean that the results are accurate. The results may be closely scattered about the wrong value!

Systematic error

Error is something wrong that needs to be put right. All too often the cause is a human mistake. Experimental errors arise from zero error, from disturbing influences such as temperature changes, and from the design of the experiment.

It is important to estimate the size, or at least the direction, of a systematic error. It's no use measuring more and more precisely if the measurements then cluster more and more closely

Measurements of speed of light

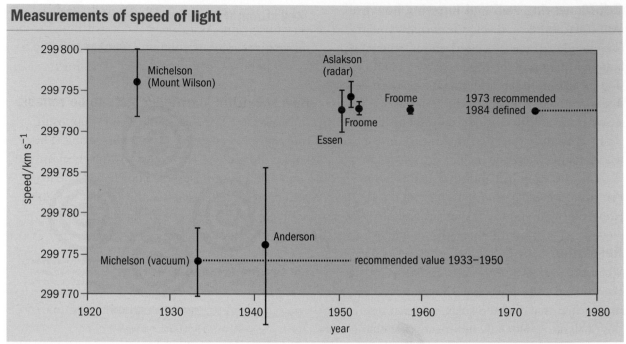

The record of measurements of the speed of light shows how the uncertainty in the measurement decreased from the 1920s to the 1970s. It also shows that the value accepted in the 1930s was 20 km s^{-1} lower than the present accepted value.

around the wrong value.

Systematic errors need to be corrected. They cannot be removed by averaging repeated results, because every result contains a similar error.

To remove systematic error requires additional or better measurements, to estimate the size of the error. The error then must be subtracted from the results. Some smart measuring devices have built-in systematic error correction – for example measuring the temperature and adjusting the results accordingly.

Case studies

The case studies at the end of this book, and on the CD-ROM, illustrate all of these ideas about measurement. They include examples of:
- achieving astonishingly high resolution and correcting for subtle systematic errors;
- the most expensive zero error in history;
- calibrating instruments for use in hospitals;
- detecting an unexpected effect through looking carefully at how results vary.

Be awkward – look for trouble

To measure as well as possible you have to find out what is going wrong – or isn't as good as it could be – and try to deal with it. For example:

- you are measuring the voltage from a sensor. Could the internal resistance be making the value too low?
- you are measuring a distance with an ultrasound rangefinder. Could the ultrasound be reflecting from the wrong object?
- you are measuring a temperature change with a thermistor. Is the thermistor correctly calibrated?

You may notice trouble when you plot results. For example, you measure the speed of a truck and the distance it runs along the floor, but the graph of distance against speed says that the truck goes a small distance at zero speed. Maybe the distance was measured from the wrong place – a systematic zero error.

These are all examples of the principle "be awkward – look for trouble". You could also call this "thinking critically". If you look for trouble you will often find it, and measure better as a result.

Look for a better way

Every experiment can be made better. If you have been looking for trouble, you will often see a way to improve things. You can also be inventive. Maybe you are finding it hard to measure the diameter of a wire precisely because it is so thin. What about weighing a length of it and using a data-book value for the density? At least this way you get a cross-check. Or perhaps you are finding it hard to see just where the image of a lamp filament is in focus on a screen. Could it be that there simply is no clear focus, because light of different colours focuses at different places? So try a colour filter. When using a stopwatch, you can get a better timing of the start if you can control it by counting down towards it.

Plot and look

Human beings see visual patterns much better than they see patterns in numbers. So when you have a number of repeated results, plot them in a simple dot-plot to see where they fall and how they spread. You'll instantly notice values that are suspiciously high or low, and ought to be investigated.

Wherever possible, as you take results of quantities that vary together (say temperature and time, or potential difference and load resistance), record them and plot a graph as you go. This will tell you, for example, when you need to take results closer together, because the graph is curving. You will also notice unusual "off-trend" values that ought to be checked out. You will also get a clear idea of how uncertain the measurements are from the way that they scatter around a trend.

Always estimate the uncertainty

A measurement with no estimate of its uncertainty is worth little. For instance, you can buy home kits to test your blood pressure. The results you get from successive readings usually vary considerably. So before getting worried about a particular measurement, you need to know how uncertain it is. There may be no need for concern at all.

Test the instruments that you use to decide just what they are capable of. How many digits in a digital voltmeter display can you trust? Just how precisely can you place a ruler and read a length measurement? This gives you an estimate of uncertainty from the properties of the instruments.

Repeat important measurements several times. Pay attention to how much they vary – it's always best to plot and look. This gives you an estimate of uncertainty from the variability of measurements.

Put these together and decide the range within which you believe the true value must lie.

The estimate of an uncertainty is itself often very uncertain. As a rule, it is not worth giving the uncertainty to more than one significant figure. When in doubt, round up rather than down.

How scientific knowledge gets to be reliable

Physicists are suspicious folk, always supposing that there is something likely to be wrong. They habitually criticise each other's results because they want to root out mistakes. They have to be like this, even though it is sometimes uncomfortable, because it's the only way for knowledge to get better and more reliable.

Measuring the speed of light

The display opposite tells a tale of decreasing uncertainty in the speed of light, but also one of unsuspected systematic error.

By 1926, the speed of light was uncertain to only about 10 parts per million. Measuring the speed between two Californian mountaintops, Albert Michelson claimed a value of $c = 299\,796 \pm 4\,\mathrm{km\,s^{-1}}$.

At this precision, the slowing of light in air (by almost $70\,\mathrm{km\,s^{-1}}$) was an important systematic error. Michelson corrected for this, but because of variations in atmospheric conditions the size of the correction needed was uncertain. He hoped to remove this uncertainty by measuring the speed in a vacuum. Results published in 1932 from measurements in a mile-long evacuated pipe gave a value about $20\,\mathrm{km\,s^{-1}}$ lower than before, being $c = 299\,774 \pm 4\,\mathrm{km\,s^{-1}}$. The new low value appeared to be confirmed by experimenters using other methods and his value was internationally adopted in 1933. However, during the Second World War, radar measurements of distance suggested that this newly accepted speed of light was too low, by about $20\,\mathrm{km\,s^{-1}}$. The error was confirmed after the war by measurements using microwaves.

Opinion among physicists shifted to the higher value. Even at the highest level in physics, unsuspected systematic error can always creep in. The origins of this error have never been fully explained. By 1973 the uncertainty had been reduced even further, to only $\pm 1\,\mathrm{m\,s^{-1}}$, and in 1984 it was agreed to define the speed of light as $c = 299\,792\,458\,\mathrm{m\,s^{-1}}$ exactly. This made the metre a derived unit, not a fundamental one.

Quick check

1. A single thermocouple gives an output of $50\,\mu V\,°C^{-1}$. If the detecting instrument can just detect a change of $0.1\,mV$ in p.d., show that the smallest change in temperature that can be resolved is $2\,°C$. If 10 such thermocouples are connected in series to make a thermopile, show that the resolution is now $0.2\,°C$.

2. Estimate the smallest change in amount of petrol in a car tank that a typical petrol gauge can show.

3. Estimate the resolution of the skin of your hand as a temperature detecting device.

4. The mains frequency is $50\,Hz$. Show that a photosensor that could detect the rise and fall of intensity from a mains lamp needs a response time of $5\,ms^{-1}$ or less.

5. Suggest two different ways to a make a sensor that produces an electrical signal that indicates the temperature of its surroundings. Compare their likely response times. Would either method be likely to require the use of
   (a) a potential divider;
   (b) an amplifier?

6. Discuss the qualities of a potential divider as a position or angle sensor under each of the headings: resolution, sensitivity, response time, noise, systematic error (see p51).

Links to the *Advancing Physics* CD-ROM

Practise with these questions:
260X Explanation–exposition *Using a sensor in a potential divider*
270D Data handling *Using non-ohmic behaviour*
280X Explanation–exposition *Filament lamp and thermistor in series*

See also in *Case studies: quality of measurement*:
200D Data handling *Calculating with uncertainties: Chapters 1–3*

Try out these activities:
180D Demonstration *Sensors reaching their limits*
200D Demonstration *Testing an infrared ear thermometer*
400E Experiment *Team sensor task briefing*

Look up these key terms in the A–Z:
Accuracy; resolution; response time; sensitivity; systematic error; uncertainty

Go further for interest by looking at:
80T Text to read *Electronic noses: telling fresh food from bad and detecting disease in cows*

Revise using the revision checklist

Summary check-up

Electricity ✓

- Electric current = $\dfrac{\text{charge transferred}}{\text{time taken}}$;

$$I = \dfrac{\Delta Q}{\Delta t}$$

- Potential difference = $\dfrac{\text{potential energy difference}}{\text{charge transferred}}$;

$$V = \dfrac{\Delta E}{\Delta Q}$$

- Conductance $G = \dfrac{I}{V}$; resistance $R = \dfrac{V}{I}$ (neither necessarily constant)
- A conductor that obeys Ohm's law has a constant conductance and resistance, so that the current is directly proportional to the potential difference
- Resistances in series add up: $R = R_1 + R_2$
- Conductances in parallel add up: $G = G_1 + G_2$, and

hence $\dfrac{1}{R} = \dfrac{1}{R_1} + \dfrac{1}{R_2}$

- The potential difference from a source, e.m.f. E and internal resistance $R_{internal}$, is $V = E - IR_{internal}$, or $V = IR_{load}$, where I is the current drawn from the source, and R_{load} is the external load resistance in the circuit
- Power in a circuit $P = IV = \dfrac{V^2}{R} = I^2R$

Sensors ✓

- Sensors can be made on a scale of micrometres, often in silicon
- Self-generating sensors include thermocouples, photovoltaic cells and piezoelectric crystals
- Modulating sensors include potential dividers, photoconductive cells, thermistors and other resistances that change with temperature, gas-sensitive materials and strain gauges
- Important qualities of a sensor are: good resolution; appropriate sensitivity; rapid response time; small unsystematic random errors, perhaps reduced by averaging; small systematic or zero error
- Sensor system design may require: dealing with lack of linearity, perhaps using look-up tables; and preventing results from being affected by other changes, notably in temperature

Circuits ✓

- Potential dividers can produce a p.d. as output from a change in resistance of a sensor
- Operational amplifiers are used in handling signals from sensors
- Circuits based on the potential divider have many uses

Questions

1. **Paper uniformly coated with graphite can be electrically conducting. A strip of the coated paper 10 mm wide and 250 mm long has a resistance of 2 kΩ.**
   (a) What is the conductance of the strip, expressed in milliamps per volt (mA V^{-1})
   (b) What would be the resistance of a strip twice as wide and twice as long?
   (c) What would be the conductance of a strip 40 mm wide and 250 mm long?
   (d) What would be the resistance of a strip 20 mm wide and 1 m long?
   (e) Suggest dimensions for a strip having resistance 500 Ω.
   (f) Suggest dimensions for a strip with conductance 1 mA V^{-1}.

2. **Give and explain an example of a potential divider used to produce an electrical output signal.**
   (a) Include in your explanation: a labelled circuit diagram showing where the output signal is obtained and where an input p.d. needs to be applied; what determines the magnitude of the output signal; what quantity the output signal indicates, and how it does so; and the effect on the output signal of attempting to measure it with a low-resistance voltmeter.
   (b) Describe a practical application for a potential divider used in the way you describe. State one difficulty that would have to be overcome in obtaining accurate results in this application.

3. **Electrons are accelerated through a p.d. of 10 kV and focused into a beam that falls on a circular spot 1 mm in diameter on a screen. The beam current is 2 mA. Calculate:**
   (a) the number of electrons falling each second on the spot;
   (b) the energy of one electron in the beam, in joules;
   (c) the power in watts delivered to the spot.
   Estimate:
   (d) the number of electrons per second striking one atom in the area of the spot.

4. **The graph below shows how the resistance of a thermistor varies with temperature.**

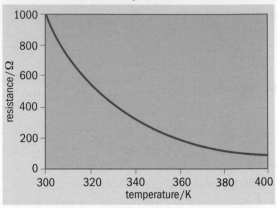

   (a) Write a caption for the graph saying what it shows and describing how the resistance varies with temperature.
   (b) Sketch a graph of how the conductance varies with temperature over the same range. Explain the shape of your graph.
   (c) If the thermistor is connected in series with a 1 kΩ fixed resistor to a source of e.m.f. 1.5 V with negligible internal resistance, (i) at what temperature would you expect the p.d. across the thermistor to be equal to that across the fixed resistor? (ii) At a certain temperature the p.d. across the thermistor is 0.5 V. What is its resistance at this temperature? Estimate the temperature. (iii) approximately what p.d. do you expect across the thermistor at a temperature of 400 K?
   (d) If the source had appreciable internal resistance, would the effects of this on the measurement be more important at high or at low temperatures? Why?

5. **Give and explain an example of each of the following:**
   (a) a sensor that has insufficient resolution;
   (b) a sensor that has too long a response time;
   (c) a sensor or measurement that has an important systematic error.
   In each case, suggest a possible way of dealing with the problem.

3 Signalling

Signals annihilate distance, bringing into contact people who are far apart. In this chapter we tell the story of the digital revolution in communication – a revolution with consequences not yet foreseen. We will give examples of:

- digital transmission of information
- uses of the electromagnetic spectrum in signalling
- calculations of limits on signalling capacity

3.1 Digital revolution: the death of distance

Following the French Revolution, the young and still disunited republic faced attacks from enemies both outside and inside its borders. Messages took days to travel across the country, so the government could not react quickly to events.

By 1793 young experimenter Claude Chappe had persuaded leaders of the revolution of the military and political advantages of building a system of semaphore telegraph stations across the country, each in sight of another, able to relay messages quickly. By 1798, government messages could travel the 400 km between Paris and Strasbourg on the German border in just 36 minutes. "By this invention, distances between places disappear...it is a means which tends to consolidate the unity of the republic," declared a member of the Committee of Public Safety. Chappe had built a national public communications network. Even today, many a hill is still called Telegraph Hill. By contrast, 100 years earlier Robert Hooke had seen semaphore as a means of private communication, as "a method for making your thoughts known far away".

Hooke and Chappe were both right. First the telephone, then e-mail and mobile phones have transformed personal communication. In place of Chappe's network of semaphore stations we now have networks of radio, cable and satellite links,

How information travels the internet

Packet switching Information sent over the internet is split into packets. Each packet has the address to which it is being sent. Internet servers decide the best route for each packet. Packets may travel by different routes, taking different times. The receiving server re-assembles the packets in sequence to reconstruct the information sent.

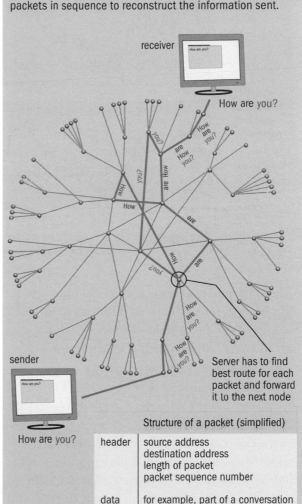

Structure of a packet (simplified)

header	source address
	destination address
	length of packet
	packet sequence number
data	for example, part of a conversation

The internet is highly interconnected, as represented by the background network shown here. Information packets can travel by many different routes from one site to another.

so that you can telephone almost anywhere in the world, send an e-mail to anyone you know, and get information stored on a computer located almost anywhere. The internet, which (just like Chappe's network for the French government) started as a US government network mainly for military communication, has developed into a worldwide

Key summary: signals and noise

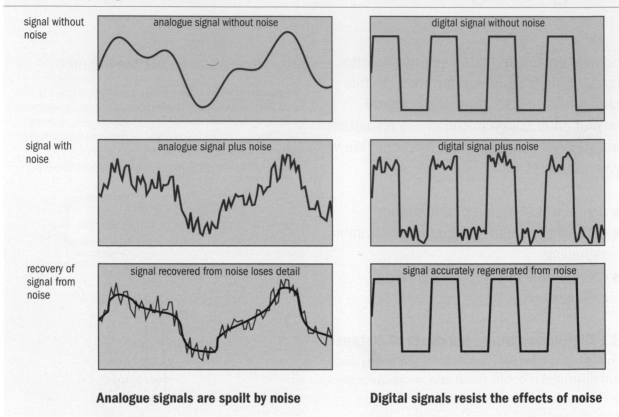

signal without noise — analogue signal without noise — digital signal without noise

signal with noise — analogue signal plus noise — digital signal plus noise

recovery of signal from noise — signal recovered from noise loses detail — signal accurately regenerated from noise

Analogue signals are spoilt by noise   **Digital signals resist the effects of noise**

medium for personal and commercial use.

It is hard to remember how recent much of this is. Worldwide radio broadcasting is less than 100 years old; national colour television broadcasting is half that age. Texting – SMS – was virtually unheard of before 1992. The consequences of the communication revolution are still working themselves out.

Not all uses of communication networks are benign. Spam messages are a plague and internet fraud is increasing. Nothing new there: in 1836 two bankers in Bordeaux were caught secretly adding news of the latest stock-market price changes to government semaphore messages.

When travelling by foot, car, train or air, distance is still important. But as far as communication is concerned, distance is effectively dead.

How will all this change people? Is it good for democracy that your vote may be swung by how well politicians have learned to perform on television? Does seeing tragic world events almost as they happen, but being helpless to do anything about them, affect people's sense of responsibility?

Will being able to contact almost anyone, almost anywhere, whenever you want, change how we think about and communicate with people?

Going digital: internet and electronic mail

The first communication systems – telephone, radio and then television – were all analogue systems. For example, in a telephone, sound vibrations were changed into matching oscillations of an electrical potential difference, which then travelled along a wire and were then changed back into sound vibrations in the air. Such analogue signalling only worked over long distances if the signal was amplified at repeater stations along the line, because the electrical oscillations got weaker as they travelled. Unfortunately the amplifiers boosted the level of any noise as much as they boosted the signal and needed careful design to avoid distortion. Going digital has changed all that, because it is easy to detect "on" and "off" signals even when weakened and noisy, and codes that correct errors can be used. A perfect copy of a message can be regenerated and sent on.

Key summary: sending a photo and sending an e-mail

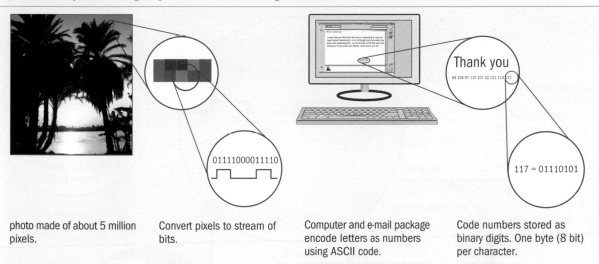

photo made of about 5 million pixels.	Convert pixels to stream of bits.	Computer and e-mail package encode letters as numbers using ASCII code.	Code numbers stored as binary digits. One byte (8 bit) per character.

Time to send a photo:

photo = 5 million pixels

1 pixel = 24 bit

broadband capacity = 10 million bits per second

$$\text{time to send a photo} = \frac{5 \text{ million pixels} \times 24 \text{ bits per pixel}}{10 \text{ million bits per second}}$$

$$= 12 \text{ s}$$

Time to send one page:

1 page = 500 words = 3000 characters approx.

= 3000 byte

= 24 000 bit

broadband capacity = 10 million bits per second

$$\text{time to send one page} = \frac{24\,000 \text{ bits per page}}{10 \text{ million bits per second}}$$

$$= 0.0024 \text{ s}$$

Run length compression:

Recode signal as runs of 1 or 0

1 1 0 0 0 0 0 0 0 0 0 0 0 0 0 0 0 1 1 1 1 1 1 1 1 0 0

2	13	8	2
of 1s	of 0s	of 1s	of 0s

Send these numbers instead.
Typical compression 9:1

More coding:

E-mail also sends data which check and correct errors.

Messages are often divided into small packets, each sent by the best route available at the moment.

Packets have to be re-assembled into messages at the receiving end.

A digital photo is made of millions of pixels, each coded for colour and intensity by a number

An e-mail is a set of numbers that code for characters and is sent as a series of 0s and 1s

---

Internet communications, e-mail (electronic mail), and mobile phones all use digital signals. In e-mail, each character is sent as a 1 byte number code (ASCII). The 256 alternatives allow enough possibilities to check errors as well as to distinguish letters of the alphabet in capital and lower case, punctuation marks, numerals, etc. A page that would take several minutes to read aloud takes a fraction of a second to send.

The rate of transmission of information available to telephone-line subscribers is rapidly increasing. Up until the year 2000, a maximum rate of $64\,000 \text{ bits}^{-1}$ was common. Now, with the increased

availability of broadband, you may have access to rates of $10 \text{ million bits}^{-1}$ or more. This means that a digital photo having several million pixels can be sent in a few seconds. The size of files that people send is growing all the time, filling the bandwidth available. This goes with the rapidly increasing size and diminishing cost of computer memory.

Rates of transmission of information are greatly speeded up by compressing the data sent. One simple way to do this is the "runs" method. For example, in a picture many of the pixels in any uniform area of the picture are the same. It is quicker to send the binary code for the number

Optics of an early compact disc system

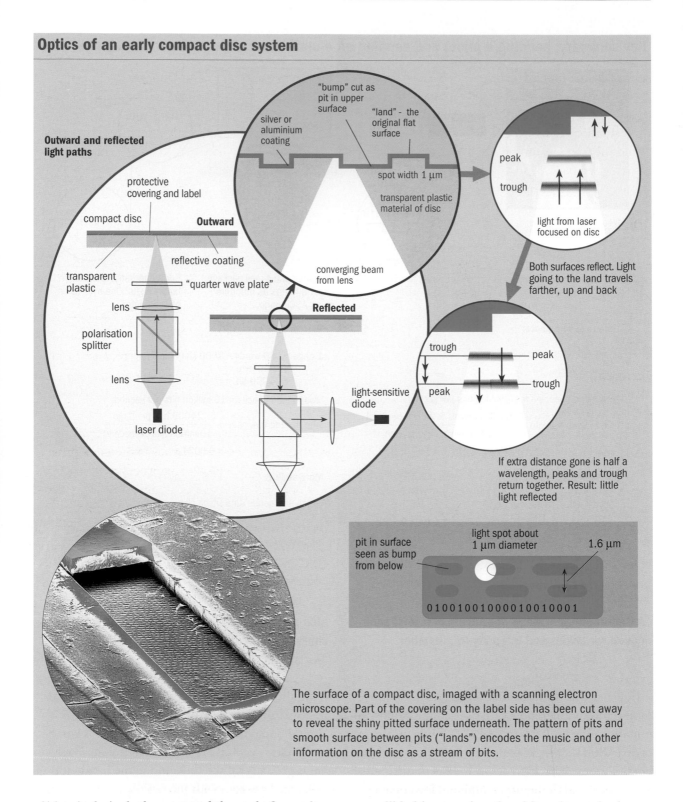

The surface of a compact disc, imaged with a scanning electron microscope. Part of the covering on the label side has been cut away to reveal the shiny pitted surface underneath. The pattern of pits and smooth surface between pits ("lands") encodes the music and other information on the disc as a stream of bits.

of identical pixels than to send the code for each and every pixel. Digital television wouldn't work without data compression (p71).

Digital music

If you play a CD or DVD, out comes top-quality stereo sound or the latest movie. The sound has no audible hiss or noise; the video picture is clear and clean. No matter that the CD or DVD may have a little dust on it, or even a scratch – the digital player works out what it should have found and plays that to you. This is what digital-signal processing and error correction can do.

A scanning electron microscope shows how the

information is stored on such discs. Tiny pits are cut into the surface of the disc in a long spiral track. In the first CDs the gap between turns was only 1.6 μm; now it is even smaller than that, and the disc may contain more than one layer of information. Lenses focus light from a laser diode onto the tiny pits on the disc's surface. The light is reflected back to a light-sensitive sensor that converts the light signal to an electrical signal.

Digitising sounds

The telephone is another application where signals are usually transmitted digitally. Digitising sound is done in three steps:

- sampling;
- binary coding;
- and further encoding.

Sampling

To digitise an image you divide it up into tiny discrete pixels and give each one a numerical value (chapter 1). How small should the pixels be? Answer: smaller than the smallest important detail in the image. To digitise a sound, you divide the varying signal into narrow time slices, or samples, and give each one a numerical value. How close together in time should the samples be? Answer: closer than the shortest time in which important changes in the signal occur, which is decided by the highest frequency that the signal contains.

In this way, a continuous signal is turned into a string of samples taken at frequent intervals. You may be surprised to learn that nothing need be lost in the process. The original can be reconstructed exactly from the samples. However, for this to be true there is an essential condition: the signal cannot contain frequencies higher than a certain maximum and the sampling must be done faster than twice that maximum frequency.

sampling frequency > 2 × highest frequency
component present

If there is an upper limit to the high frequencies present, the waveform cannot contain any unexpected wiggles between closely spaced samples. The original signal can be recovered exactly by joining up the samples with smooth

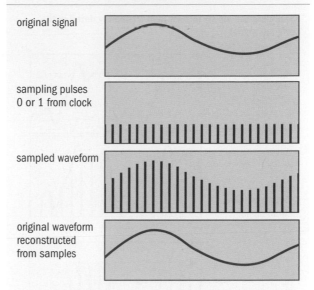

Key summary: sampling a signal

original signal

sampling pulses
0 or 1 from clock

sampled waveform

original waveform
reconstructed
from samples

The original signal can be exactly reconstructed if the sampling is frequent enough

curves of the right shape. If there are rapid wiggles, then there are higher frequencies present and the signal cannot be recovered correctly.

Worse still, such higher frequencies will generate spurious low-frequency signals called "aliases", which are not there at all. You might have seen this effect on film when the wheels on a fast-moving car seem to turn very slowly, or even backwards. The frames of film sample the scene too slowly so that the wheels make a large fraction of a turn between frames, and so appear to turn at the wrong rate.

When you are on the telephone, sometimes you don't immediately recognise a friend's voice, although you can understand what they are saying. This is because telephone transmission is restricted to the small slice of frequencies from 300 to 3400 Hz out of the whole range of frequencies you can hear, perhaps from 50 to 20 kHz. This is also the reason why music played from an answer phone sounds so terrible. Samples are taken at about 8000 times a second, a little faster than absolutely necessary.

On a CD containing music, frequencies of up to 20 kHz are needed, so the rate of sampling must be carried out at more than 40 000 times a second (44.1 kHz is standard). Filters remove frequencies of above 20 kHz, which would give trouble.

Key summary: problems with sampling

Sampling too slowly misses high frequency detail in the original signal

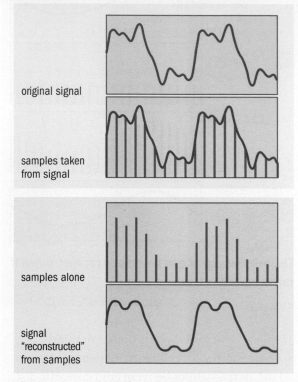

original signal

samples taken from signal

samples alone

signal "reconstructed" from samples

Sampling too slowly creates spurious low frequencies (aliases)

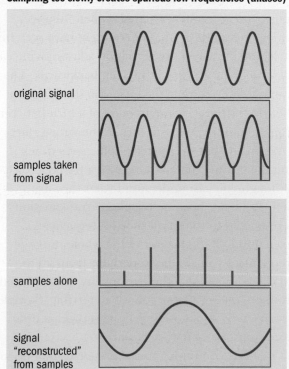

original signal

samples taken from signal

samples alone

signal "reconstructed" from samples

Sampling must be done frequently enough to reconstruct the original signal

Key summary: digitising a signal

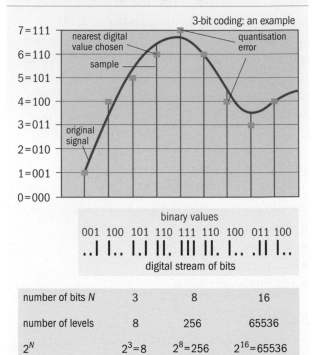

3-bit coding: an example

nearest digital value chosen

quantisation error

sample

original signal

number of bits N	3	8	16
number of levels	8	256	65536
2^N	$2^3=8$	$2^8=256$	$2^{16}=65536$

The greater the number of bits the better the resolution

Binary coding

Look back to the antenatal scan on p1. Each pixel in the image is assigned one of 256 levels of grey, using 8 bits or 1 byte of information per pixel. Sound sampling for telephony uses the same scale. An 8-bit analogue-to-digital converter turns each sample strength into a number from 0 to 255 on the binary scale. This is quite coarse, but good enough for intelligible speech. Higher quality CD sound requires better resolution: it uses 16 bits, giving 65 536 different levels for each sample.

The telephone signal has now been converted into a stream of bits, a bit stream. Samples come in blocks of eight, each block giving the magnitude of one sample. Obviously, the 8 bits for one sample must be sent before it is time for the next sample to come along 1/8000 of a second later. So the bits must be sent at a rate of at least 64 000 bits s^{-1}.

CD-quality sound demands a higher channel capacity. 16-bit samples must be sent at a rate of 44.1 kHz, so the bits must come along 16 times as fast, at 0.7 MHz. Squeezing in two stereo channels doubles the rate to 1.4 MHz, a frequency similar to

Key summary: samples and strings of bits

sampling stereo sound for a compact disc

sampling rate
minimum sampling rate = 2 × maximum frequency
maximum frequency = 20 kHz
actual sampling rate = 44.1 kHz
time between samples = 22.7 μs

rate to send bits
32 bits every 22.7 μs
rate of sending bits = 32 bit/22.7 μs
rate of sending bits = 1.4 million bits per second
1.4 MHz

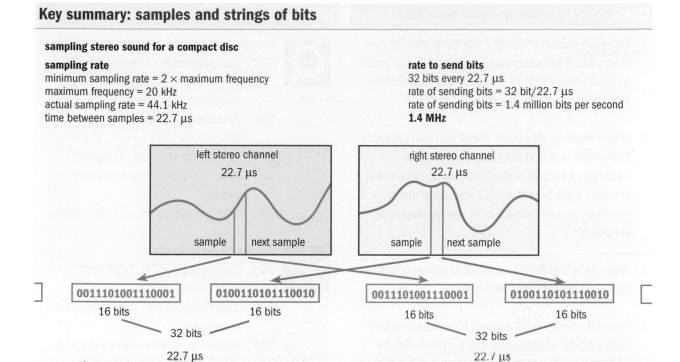

High quality means rapid sampling of many bits

those in the medium radio waveband (see p65).

In general, the channel capacity needed to transmit a digitally sampled signal depends on:

- the resolution of each sample, that is the number of bits specifying the value of each sample;
- and the rate of sampling, which must be at least twice the highest frequency in the signal spectrum.

Further encoding

People take it for granted that e-mail will arrive without errors, and expect CD sound and DVD images to be fault-free. In fact, any transmission of a signal introduces errors; a few bits get lost or changed because of noise. In spite of that, digital-signal transmissions are remarkably free of errors. This is achieved by coding the signals in such a way that errors can be located and corrected. Actual error-correction codes are complicated, but are based on a simple idea: sending more information than the minimum. This is called adding redundancy. A crude way to do it would be to send each part of the signal more than once. If the copies do not agree there must be an error and the signals can be asked for again until they agree. Much

cleverer ways than this are used in practice, though making repeated readings is used in bar-code readers at supermarket tills. The essential point is that using digital signalling the system computes what it should have received. A CD is so tolerant that you can drill a 2 mm-diameter hole in it and barely notice an effect on the sound it produces.

Digital futures

A consequence of the fact that digital information is just numbers that can be used in a computation is that different kinds of information can be combined: e-mails can have images and tunes attached, for example. Digital signals, being numbers, are also easy to encrypt or scramble. This is going to be a big issue in the future: who will have the right to code and decode different signals?

As electronic commerce develops, will it be possible to keep your credit-card number secure over the internet? What kinds of materials will you have delivered to you as electronic files: newspapers, books, music, brochures, videos, pictures to put on your wall? Can ownership of such information be protected? Should it be? The digital future poses many such problems.

Quick check

1. Suppose you chat using a digital mobile phone that uses 8-bit sampling sent 8000 times each second. Show that in half an hour 115 million bits of information are transmitted.

2. When reading this book aloud you can probably cover 200 words a minute. Assuming an average of six characters per word, show that it would take less than 0.2 s to send these words by e-mail using 1 byte per character at $64\,000\,\text{bit}\,\text{s}^{-1}$.

3. Show that the ASCII code 32 is represented by the 8-bit binary code 00100000.

4. Show that nearly 80 million bits must be read using 16-bit sampling to get just one minute of high quality stereo music from a CD if the maximum frequency is 20 kHz.

5. The recording area of a CD measures 35 mm from inside to outside, and has an average radius of 40 mm. The recording track spirals with a spacing of 1.6 µm. Show that
   (a) the spiral makes about 22 000 turns;
   (b) the total length of the spiral is over 5 km.
   If the track is scanned at $1.25\,\text{m}\,\text{s}^{-1}$, show that the total playing time is over one hour.

6. Make a sketch to show how sampling a signal too slowly can introduce frequencies that are not present in the original signal.

Links to the *Advancing Physics* CD-ROM

Practise with these questions:

10C Comprehension *Teleworking: working from home using telecommunications*

10E Estimate *Making estimates about information*

50S Short answer *Simple sampling*

60S Short answer *Sampling repetitive motion*

70S Short answer *Sampling and hearing*

Try out these activities:

10S Software-based *Data on the telecommunications explosion*

50E Experiment *Guess a waveform from a sample*

70S Software-based *Looking less often*

Look up these key terms in the A–Z:
Amount and rate of transmission of information; analogue to digital conversion; bandwidth; bits and bytes; digital sampling

Go further for interest by looking at:

10T Text to read *Semaphore unites the new republic*

Revise using the revision checklist and:

100 OHT *Seeing sampling*

200 OHT *Signal sampling*

300 OHT *Problems with digital sampling*

400 OHT *Digitising samples*

500 OHT *Effect of noise on analogue and digital signals*

3.2 Signalling with electromagnetic waves

Spin the tuning dial of a radio (or press the seek button) and you will hear, one after the other, different stations at specific frequencies. The radio sits in a space filled with signals from all of the stations, all on top of one another. To hear one station the radio must be tuned to select just that one frequency. If you mapped out all of the radio frequencies on a scale, with a line at each frequency, making the line longer the stronger the signal, you would produce a spectrum of the radio frequencies that your set can pick up.

Imagine going for a swim in the sea. Long high waves lift your body and let it drop. Shorter waves splash against your face. Tiny ripples pass unnoticed. You are swimming in a spectrum of water waves, all just movements of the water surface happening at once. Now imagine standing outside in the sunshine. Rapid electromagnetic oscillations at infrared frequencies fall on your skin and warm you. Electromagnetic waves you don't feel pass around you: you know they are there when you turn the radio on. A mobile phone rings: more radio waves are sent and received. You are swimming in a sea of electromagnetic waves, all just oscillations of electric and magnetic fields in empty space, all on top of one another.

Swimming in a sea of electromagnetic waves

Aerials and wavelengths

reflector

vertical rods for vertical polarisation

receiving rods

approximately half a wavelength

Key summary: communication wavebands

frequency			wavelength
30 kHz	LF low frequency	navigation, radio beacons, long-distance broadcasting	10 km
300 kHz	MF medium frequency	national broadcasting, aeronautical navigation	1 km
3 MHz	HF high frequency	long-distance broadcasting, amateur radio, maritime radio	100 m
30 MHz	VHF very high frequency	FM radio, mobile radio communications	10 m
300 MHz	UHF ultra high frequency	television, mobile telephone networks	1 m
3 GHz	SHF super high frequency	satellite links, ground microwave links, radar	100 mm
30 GHz	EHF extremely high frequency	radar, radio astronomy	10 mm
300 GHz	far infrared	infrared astronomy	1 mm
3 THz	mid infrared	infrared astronomy	100 μm
30 THz	near infrared	optical fibre, remote controls, bar codes, CD player	10 μm
300 THz			1 μm

Communications use wavelengths of between 1 μm and 10 km

Key summary: electromagnetic waves from an aerial

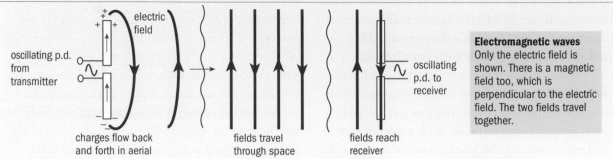

Electromagnetic waves
Only the electric field is shown. There is a magnetic field too, which is perpendicular to the electric field. The two fields travel together.

Electromagnetic waves are transverse waves. A rod aerial lies parallel to the electric field.

Key summary: polarisation by scattering

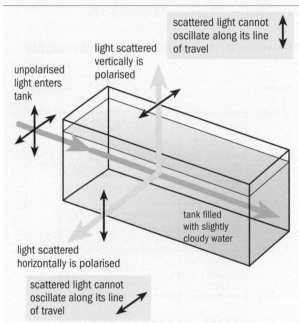

Light from a blue sky is polarised by scattering

When radio broadcasting in Britain first began in the early 1920s, a single 1500 m wavelength signal, frequency 200 kHz, covered most of the country, its long wavelength meaning that hills hardly got in the way (see chapter 6). Today, frequencies used for communication stretch up to those of the infrared waves used in optical fibres. In between are the microwaves used for telephone links overland and to satellites.

Looking at radio and television aerials on the roof of a house can give you an idea of the size of wavelengths being picked up. Typically, aerial rods are half a wavelength long. Television and mobile phones use the ultrahigh frequency (UHF) band

300 MHz–3 GHz. The wavelength is a few tenths of a metre. FM radio uses the VHF band from 30 to 300 MHz, with wavelengths of a few metres. Satellite transmissions use microwaves with wavelengths of a few tens of millimetres; the aerial is a collecting horn at the focus of a dish.

Polarisation

You can tell something else about a signal by looking at an aerial on a roof: its direction of polarisation. Radio waves make electrons surge up and down aerial rods at the frequency of transmission. The transmitter sent the radio waves by making currents oscillate in a similar rod. The varying electric field in the wave points parallel to the rods and parallel to the direction of motion of the electrons. To pick up the signal, the receiving aerial must be parallel to the changing electric field.

In general radio signals are polarised either vertically or horizontally. On my roof, the FM radio aerial is aligned for horizontal polarisation, and the TV aerial for vertical polarisation.

A sound wave cannot be polarised like this because the oscillations in the wave are along the direction of wave propagation: the wave is longitudinal. A radio wave can be polarised because the oscillating electric and magnetic fields are at right angles to the direction of propagation: the wave is transverse.

If you go fishing or skiing you might wear Polaroid spectacles. They cut out visible light that is polarised in one direction, such as light reflected from water. Light scattered high in the atmosphere, making the sky look blue, is also polarised, and some insects use this direction of polarisation to navigate their way from nest to plant.

CD systems also make use of polarised light. Light is steered through the system by a polarising prism that transmits light polarised in one direction and reflects light polarised in the other direction.

What's in a signal?

Press two notes on a piano or pluck two strings on a guitar. Your ear takes the complex resulting vibration and resolves it into two notes – the spectrum of the sound. This is why musicians can learn to identify notes in a chord. Your eye does not do the same for light, however. A mixture of red and blue looks purple, not red plus blue. Computer programs can take a repeating signal and analyse it to see what frequencies it contains. Your ear can do it; your eye cannot. A prism does it for light and a tunable radio set does it for radio waves. If signals are digitised, computer analysis can be done quickly: you can see the spectrum of a sound as you make it. The analysis (Fourier analysis) is like a mathematical tuning dial, picking out frequency components one by one.

Think for a moment about hearing impairment. One kind makes the ear insensitive to high frequencies (this often happens as you grow older). The high-frequency components in sounds such as "s" and "t" become inaudible. "Bit" might be hard to tell from "bid". Tapes of sounds filtered to be like those heard by people with various impairments can help you to understand their problems.

The examples above all show that the information in a signal is contained in the full range of the frequencies in its spectrum. If some of the frequencies are lost, then so is some of the information in the signal. In fact, the spectrum of a repeating waveform contains all of the information of the original waveform. Each can be reconstructed from the other.

If you play a synthesizer you make musical tones by mixing the required frequencies: you go from spectrum to sound. When you identify a chord or use a computer program to analyse a sound you are going the other way, from sound to spectrum.

Sending many signals at once

If you call a friend in another country, your call is transmitted as 8-bit samples, with one sample sent

Waveforms and spectra of sounds

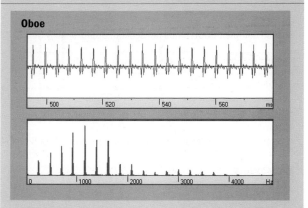

The spectrum of an oboe contains a number of distinct frequencies spaced in simple relationship to one another, and extends over a range of frequencies

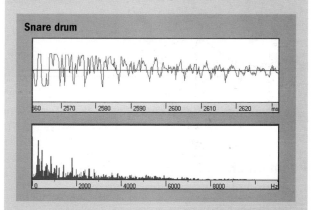

The spectrum of a drum beat spreads widely over a range of frequencies

every $1/8000\,s$ (roughly every $100\,\mu s$). The bits can be sent by changing the amplitude (or frequency) of a high-frequency wave backwards and forwards between two values. Obviously the wave cannot be switched faster than the wave oscillates. However, if the signal goes by a microwave link of $10\,GHz$ frequency, the microwave carrier makes a million oscillations in the $100\,\mu s$ gap between your samples. Into that gap can be fitted samples from many other callers if each is given a private time slot. By using a high transmission frequency, many telephone channels can be fitted into one radio-frequency band. Optical fibres, which use much higher infrared frequencies, are able to carry many more bits per second; by 1990 it was $2\,Gbit\,s^{-1}$ – room for approximately $30\,000$ telephone channels.

Key summary: time-division multiplexing – sending many phone calls down the same pipe

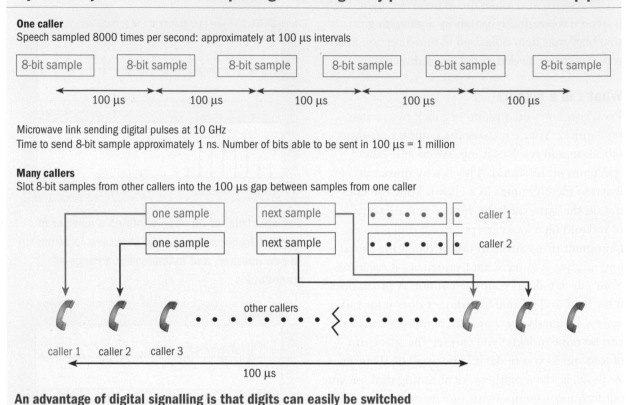

One caller

Speech sampled 8000 times per second: approximately at 100 μs intervals

| 8-bit sample | 8-bit sample | 8-bit sample | 8-bit sample | 8-bit sample | 8-bit sample |

       100 μs   100 μs   100 μs   100 μs   100 μs

Microwave link sending digital pulses at 10 GHz
Time to send 8-bit sample approximately 1 ns. Number of bits able to be sent in 100 μs = 1 million

Many callers

Slot 8-bit samples from other callers into the 100 μs gap between samples from one caller

| one sample | next sample | • • • • • | • | caller 1 |
| one sample | next sample | • • • • • | • | caller 2 |

other callers

caller 1   caller 2   caller 3

100 μs

An advantage of digital signalling is that digits can easily be switched

Room on air

Each kind of signal contains a range of frequencies, known as the spectrum of the signal. Telephone conversations have a spectrum width of 3100 Hz (300 to 3400 Hz). High-fidelity music has a spectrum width of 20 kHz.

You may have noticed how the frequencies of FM radio stations are spaced out. One may be 98.4 MHz; the next nearest will never be closer than 98.2 MHz or 98.6 MHz, always at least 0.2 MHz (200 kHz) apart. This is because sending a signal on a radio carrier spreads out the frequencies in the spectrum of the carrier.

Radio, television, mobile phones, satellite and microwave links, all use different parts of the electromagnetic spectrum (see p65). Any such channel can only transmit a signal if it can transmit a sufficiently large range of frequencies. The range of frequencies that a channel can handle is called its bandwidth. The faster that information needs to be transmitted, then the larger the bandwidth and the higher the frequency band needed. The bandwidth required for a signal

limits the number of stations that can be fitted into a given range of frequencies or "waveband".

Bandwidth for sale

The communications revolution has produced a new and valuable commodity for sale – bandwidth. Mobile-phone companies have paid governments for the right to use ranges of frequencies, and now they need to create services that we all want to buy, so that they can make a profit on that investment.

Why then is bandwidth so valuable? The reason is that the bandwidth of a channel – the range of frequencies used – determines the rate at which information can be sent along that channel. So mobile-phone firms and internet service providers are selling you the ability to send information quickly. The monthly rate you pay for a mobile-phone contract increases with the amount of information-carrying capacity you get. Similarly, your broadband internet connection may promise 10 or even 100 million bits per second. If you are (or were) restricted to a telephone dial-up connection, the rate on offer is only 64 000 bits^{-1}.

Key summary: bandwidth for digital signals

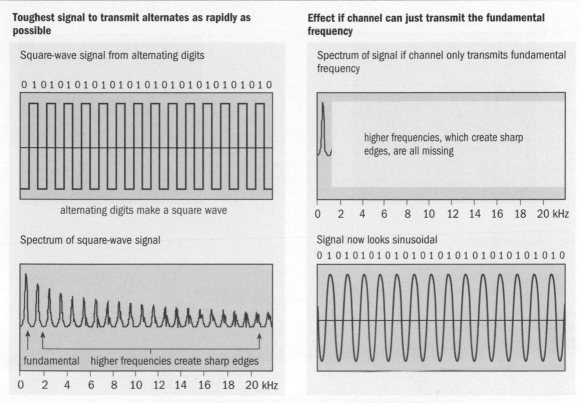

Toughest signal to transmit alternates as rapidly as possible

Square-wave signal from alternating digits

0 1 0 1 0 1 0 1 0 1 0 1 0 1 0 1 0 1 0 1 0 1 0 1 0 1 0 1 0 1 0 1 0

alternating digits make a square wave

Spectrum of square-wave signal

fundamental   higher frequencies create sharp edges

0   2   4   6   8   10   12   14   16   18   20 kHz

Effect if channel can just transmit the fundamental frequency

Spectrum of signal if channel only transmits fundamental frequency

higher frequencies, which create sharp edges, are all missing

0   2   4   6   8   10   12   14   16   18   20 kHz

Signal now looks sinusoidal

0 1 0 1 0 1 0 1 0 1 0 1 0 1 0 1 0 1 0 1 0 1 0 1 0 1 0 1 0 1 0 1 0

Original bits can be recovered if the channel bandwidth is more than half the bit rate

Bandwidth and information-carrying capacity

Suppose you want to send a digital signal at a rate of b bits s^{-1} along a given channel, say a telephone line. How much bandwidth do you need?

The largest signal frequency would be if you were to send alternating binary digits: 010101010101. The channel must be able to transmit the frequency of these fast-repeating pairs. If it can't, then it is unable to respond quickly enough to transmit this worst-case signal. If it can, it can see the bits, even though they now look like round-shouldered sinusoidal pulses. That is enough to regenerate clean new bits to send on. So the bandwidth must at least cover this frequency range.

In this limit, a pair of bits "01" looks like one cycle of a wave.

$$\text{bandwidth } B \text{ needed} = \frac{b \text{ bits per second}}{2}$$

For practical purposes the bandwidth required is roughly equal to the rate at which bits must be sent.

This is not the only requirement. The noise level

(unwanted variations superimposed on the signal) in the channel must be small compared with the signal strength so that the bits are not lost in noise. Noise can be tolerated as long as it is possible to tell the high level of a "1" from the low level of a "0". The digital signal can then be reconstituted and the effect of noise removed.

Bandwidth needed to go digital

Communication systems are shifting rapidly from analogue to digital transmission, with plans to turn off the terrestrial analogue TV signal. What demands does this place on the bandwidth?

This question was answered in 1949 by communications engineer Claude Shannon. He worked out how many bits per second are required to send a digitised version of an analogue signal.

The signal must be sampled at twice the highest frequency it contains. If the width of the spectrum of the signal is W, then:

$$\text{number of samples per second} = 2W$$

Key summary: signal, noise and bandwidth

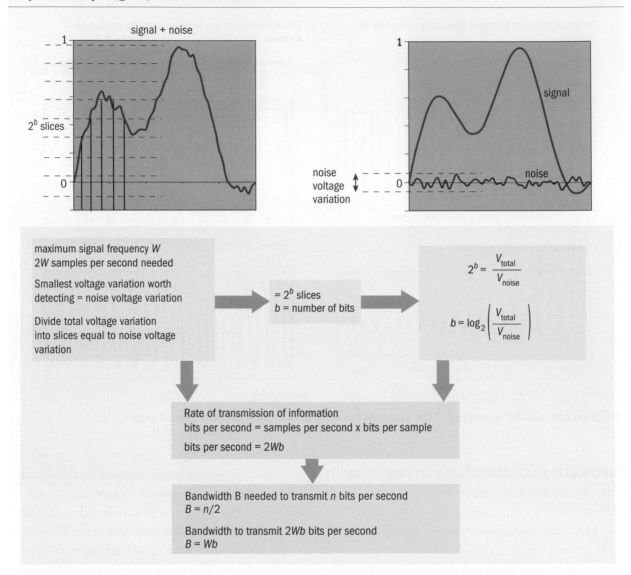

signal + noise

2^b slices

1

0

signal

noise voltage variation

1

noise

0

maximum signal frequency W
$2W$ samples per second needed

Smallest voltage variation worth
detecting = noise voltage variation

Divide total voltage variation
into slices equal to noise voltage
variation

= 2^b slices
b = number of bits

$$2^b = \frac{V_{total}}{V_{noise}}$$

$$b = \log_2\left(\frac{V_{total}}{V_{noise}}\right)$$

Rate of transmission of information
bits per second = samples per second x bits per sample

bits per second = $2Wb$

Bandwidth B needed to transmit n bits per second
$B = n/2$

Bandwidth to transmit $2Wb$ bits per second
$B = Wb$

Rate of transmission of information increases with bandwidth. Noise decreases the possible rate of transmission of information.

If b bits per sample are needed then:

rate of transmission of information =
bits per second =
samples per second $\times$ bits per sample

$$rate = 2Wb$$

The bandwidth B required is then just half the rate of transmission of information:

$$B = Wb$$

You can see that digital transmission needs more bandwidth than analogue transmission. The gain is that the signal is transmitted with less error, leading to better TV pictures, for example.

It may look as if you can choose any number b of bits per sample, but this is not the case. The reason is the effect of noise. As you increase the number of bits per sample, you increase the number of slices into which the signal voltage variation is divided. b bits gives 2^b slices. If the slices become smaller than the voltage variations due to noise, the bits generated become

Digital television bandwidth

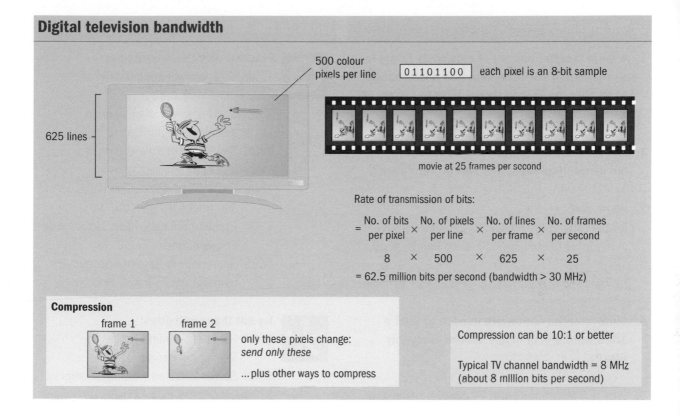

Compression

only these pixels change:
send only these

... plus other ways to compress

Compression can be 10:1 or better

Typical TV channel bandwidth = 8 MHz
(about 8 million bits per second)

---

meaningless. Thus noise sets an upper limit on the number of bits per sample.

$$2^b \le \frac{\text{total noisy signal variation}}{\text{noise variation}} = \frac{V_{\text{total}}}{V_{\text{noise}}}$$

The largest number of bits per sample worth using is therefore:

$$b = \log_2 \frac{V_{\text{total}}}{V_{\text{noise}}}$$

The amount of noise on a channel limits the bandwidth it can offer. In fact, you can't get a broadband telephone line connection if you live more than a few kilometres from the nearest telephone exchange. If the telephone line is long, the signal is weak and the noise is large, so the bandwidth available is reduced. So some places in the UK will have to depend on satellites or on fibre-optic cable instead.

Communications engineers measure the ratio of signal power S to noise power N. Rewritten in terms of these quantities the number of bits per sample is:

$$b = \tfrac{1}{2}\log_2\left(1 + \frac{S}{N}\right)$$

Digital television

Television signals gobble up bandwidth, typically 8 MHz for each channel. Space in the UHF band is limited. But a satellite broadcasting at microwave frequencies of up to 30 GHz may have a bandwidth of say 1 GHz, which is enough for more than 100 television channels. It is clear that the problem now is not providing more television channels, but finding the people, time and money to put something worth watching on them all.

At first sight, digital television demands even more bandwidth. The huge number of bits needing to be sent has to be cut by compressing the signal. One method is to send after each frame only the pixels that change to give the next frame. Many frames have large areas of picture in common, so this can be very economical. If a frame doesn't change, nothing needs to be sent!

Of all the ideas in this chapter, perhaps the most important is the twin pair of a waveform and its spectrum – the picture in time and the picture in frequencies. This idea, under the name Fourier analysis, appears again and again in physics – in optics, in quantum theory and in many other places.

Quick check

1. A radio receiver has aerial rods each 75 mm long so that a pair correspond to half a wavelength. Show that they are designed for a frequency of 1 GHz.

2. The radio emission from some galaxies is polarised. How could this fact be confirmed?

3. Show that
   (a) the ratio of the highest to the lowest frequency in the table of the LF to EHF radio bands on p65 is 10 million;
   (b) the ratio of their wavelengths is also 10 million.

4. Show that the bandwidth needed to send a 1000 word e-mail in one second is roughly 60 kHz.

5. Why is 8 MHz bandwidth TV never transmitted using the medium frequency wave band? (See the table on p65.)

6. A JPEG image contains about 4 million pixels, each of one bit. Show that if it could be sent using FM radio with a bandwidth of 200 kHz it would take 20 s to be transmitted.

Links to the *Advancing Physics* CD-ROM

Practise with these questions:

120S Short answer *Longitudinal and transverse waves*

130S Short answer *Polarisation in satellite communication*

140S Short answer *Polarisation in practice*

150X Explanation–exposition *Waveforms and frequency spectra*

170X Explanation–exposition *Music on an answerphone*

150S Short answer *Rate of transmission of information limited by noise*

Try out these activities:

150H Home experiment *Telling frequencies apart*

160S Software-based *Filtering sounds*

210S Software-based *Cleaning up a sound*

220S Software-based *Building up a sound*

260S Software-based *Bits per second and bandwidth*

Look up these key terms in the A–Z:
Amount and rate of transmission of information; bandwidth; digital sampling; electromagnetic spectrum; noise; polarisation; radio waves; spectrum of a signal

Go further for interest by looking at:

30T Text to read *Digital recording error correction*

100D Demonstration *The CD with the hole*

Revise using the revision checklist and:

800 OHT *Signal bands for communication*

1200 OHT *Signal, noise and bandwidth*

Summary check-up

Signals ✓

- Signals are either digital or analogue. Digital signals are being used more and more.
- Digital signals have the advantage that: they can be reproduced and retransmitted easily; they minimise the effects of noise; they can be processed and encoded by computer; they can represent different kinds of information in a uniform way

Sampling and encoding ✓

- Analogue signals are digitised by sampling
- Sampling must be done at a rate greater than twice the bandwidth (often, twice the highest frequency present)
- Samples can be read with different resolution (e.g. 8 bit, 16 bit). N bits gives 2^N different levels of measured values
- Errors can be detected and removed if the signal is suitably encoded

Polarisation ✓

- Electromagnetic waves, being transverse waves, can be polarised

Bandwidth, spectrum, transmission rate ✓

- Communication with electromagnetic waves uses frequencies from a few thousand hertz to infrared frequencies and above, divided into bands used for different purposes
- A signal can be analysed into the frequencies it consists of, i.e. its spectrum
- A signal channel has a capacity (the rate at which it can transmit information) measured in bits per second
- The bandwidth of a channel is the range of frequencies it can transmit
- The bandwidth needed to transmit a digital signal is of the order of magnitude of the maximum number of bits to be sent per second
- Noise in an analogue signal limits the number of bits per sample that it is worth using

Questions

1. The spiral track of a music CD plays for one hour, read at $1.25\,\text{m s}^{-1}$. The two-channel stereo music on the CD has frequencies extending up to 20 kHz. The sampling frequency on each channel is 44.1 kHz. Samples are digitised with 16 bits per sample.
   (a) Show that the track is 4.5 km long.
   (b) Give a reason why the sampling frequency is more than 40 kHz.
   (c) How many bits does the player have to read per second for each stereo channel?
   (d) Estimate the bandwidth of the communication between reading head and player.
   (e) Show that there are 5×10^9 bits for one hour of stereo music.
   (f) How much track length is there per bit? Comment on this.

2. The table below shows how the numbers of transatlantic and transpacific telephone lines have changed over time. Plot appropriate charts to display the changes and to compare cable and satellite. Suggest reasons for the changes in capacity that have occurred. (Consider whether or not to use a logarithmic scale.)

Numbers of telephone lines*

	Transatlantic		Transpacific	
Year	Cable	Satellite	Cable	Satellite
1986	22 000	78 000	2 000	39 000
1987	22 000	78 000	38 000	39 000
1988	60 000	78 000	38 000	39 000
1989	145 000	93 000	38 000	39 000
1990	145 000	283 000	38 000	39 000
1991	221 000	283 000	114 000	27 000
1992	296 000	496 000	190 000	27 000
1993	410 000	621 000	264 000	83 000
1994	701 000	621 000	264 000	234 000
1995	1 311 000	711 000	264 000	234 000
1996	1 311 000	711 000	864 000	234 000
1997–2000 (estimate)	1 311 000	738 000	1 465 000	424 000

Source: *Cairncross F The Death of Distance (1997) Orion Business Books. Sourced there to TeleGeography, Inc. (Washington, DC).

3. There is a photograph on the Web that I want to download. It is 30 Mbyte in size. I do not want to wait more than half a minute for it to arrive. The picture is in colour, with 8 bits per pixel per colour.
   (a) What rate of transmission in bits per second do I need the Web connection to provide?
   (b) Can this rate be achieved down an internet connection of bandwidth $1\,\text{Mbit s}^{-1}$?
   (c) How many pixels should I expect the picture to have? If I print it 100 mm square, how big will a pixel be?

4. Mobile phones use the frequency range 800–900 MHz. Speech frequencies of up to 4 kHz are transmitted.
   (a) What is the range of radio wavelengths used?
   (b) At what rate must the speech be sampled?
   (c) If samples are digitised with 1 byte per sample, how many distinct levels of signal can be sent?
   (d) What rate of transmission in bits per second must the line achieve?
   (e) Estimate the bandwidth used by one line.
   (f) Estimate how many telephone lines can be handled in the 100 MHz band between 800 and 900 MHz.

5. Describe evidence for the polarisation of:
   (a) microwaves;
   (b) visible light.
   Suggest a practical application for the polarisation of one of them.

6. A noisy analogue signal has a total voltage variation (signal and noise) of 512 mV. The signal is to be sampled with 8 bits per sample, at a rate of 40×10^3 samples per second.
   (a) What is the largest noise voltage variation that can be tolerated?
   (b) What is the highest frequency that can be present in the noisy signal?

4 Testing materials

Here we begin the story of materials in use, their properties and how these properties depend on the underlying structure of the material. We will give examples of:

- different types of materials
- testing materials and properties
- various uses of materials

Using a sharp chromium-steel blade, the surgeon carefully slices into your eyeball. A fine tube sucks out the jelly-like contents of the lens. A clear polymer disc is inserted into the lens sac and you can see again! This is now an everyday operation for people suffering from cataracts, but it wouldn't be possible without the materials and techniques that have been developed in the last few years.

Much of our way of life – our communications and transport, health, sport and leisure, and the progress of science itself – depends on modern materials. Most of the time we take this for granted and just get on with our lives. But all the time people are designing materials to satisfy ever greater demands: for taller buildings, faster trains, smaller computers, longer-lasting body implants, a cleaner world.

This is a part of science where, every day, people are making new discoveries and producing new materials that have a direct impact on our lives.

4.1 Making the best choice

For some jobs you need just the right material with the right combination of properties. Here we look at some interesting objects and the materials, both natural and synthetic, from which they are made.

Superconductors

In a body scanner (above, right) a patient lies surrounded by a powerful electromagnet. The magnet is a coil of superconducting wire. When cooled to within four degrees of absolute zero, the wire loses all electrical resistance. A current can then flow through the coil wasting no electrical energy, and a strong magnetic field is produced.

Coalbrookdale by night (1801) – an oil painting by Philippe Jacques de Loutherbough (1740–1812) – shows one of the Coalbrookdale ironworks silhouetted against the fiery glow of a furnace being tapped of its ore. The development of coke smelting in this area of Shropshire, by Abraham Darby in the 18th century, revolutionised iron production and helped fuel the Industrial Revolution. The area's unique combination of natural resources led the ironworks to produce Britain's first iron rails, iron bridge, iron boat and steam locomotive.

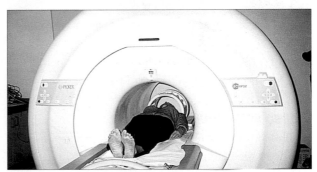

Magnetic resonance imaging is a non-invasive technique used to render images of the inside of a patient's body.

Fibre-reinforced polymers

The chassis of a racing car is made from plastic strengthened with carbon fibres, making it lighter and stronger than a metal chassis. Fibre-reinforced polymers can be used to make complex structures. A new design can take less than two weeks to get from a designer's screen to a Grand Prix race track.

Yew wood

An ancient spear (p76, top right) was found at Clacton in Essex. At 250 000 years old, it is one of the oldest wooden artefacts ever found. Yew wood is extremely hard and when sharpened to a point it would have been capable of penetrating the thick

This racing car chassis is made from plastic strengthened with carbon fibres, and is lighter and stronger than a metal one.

This Yew wood spear is extremely hard. When sharpened to a point it would be capable of penetrating animal hide.

Solar panels provide a clean, free-to-run source of electricity.

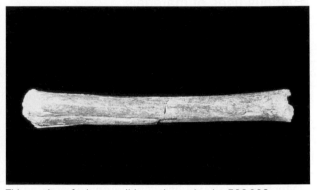

This portion of a human tibia, estimated to be 500 000 years old, was discovered by archaeologists at Boxgrove in 1993.

hide of a bison or even a rhinoceros. Its hardness also helped to preserve it down the millennia.

Silicon solar cells

A solar battery can provide a clean, free-to-run source of electricity in a remote area. It can power a pharmacy's fridge for storing medicines, or a communal television set. Silicon is a semiconductor; some of the electrons in silicon are only weakly bound to atoms, and the energy of light is enough to set them free to carry an electric current.

An ancient bone

Scientists and engineers aren't the only people who need to know about materials. What follows is the story of some ancient materials and the archaeologists who struggled to understand them.

In 1993 archaeologists working at Boxgrove in Sussex found part of a human bone that was estimated to be half a million years old. It was half of a tibia – one of the bones in the lower leg. Later, they found a single tooth. Alongside these finds they uncovered many tools and the remains of

butchered animals. But what can you say about a group of people when all you have is one tooth and half a leg bone?

The femur, the single bone in the thigh, is the thickest bone in your body. It has to bear the weight of your body pressing down from above. And when you run, there is a large upwards force on the femur every time your foot hits the ground, perhaps three times your body weight if you are sprinting. So the femur must be strong in compression. If you do a lot of running around, your body responds by building even thicker, stronger bones.

Detailed measurements of the Boxgrove tibia showed that it was remarkably thick and strong, much more so than the tibia of a modern human. This suggested that it came from a tall, powerful individual who led a very active hunting life.

The stuff of skeletons

Bone is a remarkable substance. It is a **composite material**, made from two very different substances:
- crystals of calcium phosphate, a brittle ceramic material, strong in compression, rather like chalk;

Key summary: breaking bones

Bone in compression
Compressive forces squash the bone.

Bone in tension
Tensile forces stretch the bone along its length.

thick thigh bone

thinner arm bone

Bone bends
On the outside of the bend, bone is in tension. On the inside of the bend, bone is in compression.

Bone breaks
A break occurs where a crack develops in the surface that is in tension.

Brittle materials are generally stronger in compression than tension

Bone: a composite material

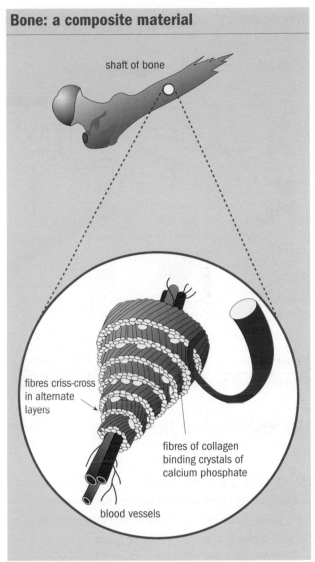

shaft of bone

fibres criss-cross in alternate layers

fibres of collagen binding crystals of calcium phosphate

blood vessels

Bone is composed of two materials. The calcium phosphate crystals are embedded in a mass of protective, slightly flexible collagen fibres. For added strength, the fibres, which are arranged in alternate layers, are packed in a criss-cross fashion.

● and fibres of collagen, a polymer made of protein, strong in tension, rather like rubber. These materials belong to two different classes: ceramics and polymers. **Ceramic materials** are hard and brittle; they include china and other pottery, as well as more modern "engineering ceramics", such as alumina and silicon carbide. **Polymers** include the familiar synthetic materials we often call plastics, such as polythene and polyester, as well as many natural materials such as cotton and leather.

Bone is nature's equivalent of modern carbon-fibre-reinforced plastic, a tough, strong but lightweight material. Composite materials are designed to combine the most desirable properties of two or more different materials.

Tough tools

The Boxgrove archaeologists found that the people who lived there half a million years ago used a variety of tools. These people knapped flint to make sharp knives and hand axes, which is highly skilled work. You need to know just

where to strike a lump of flint to remove flakes so as to leave a useful core. Flint is a mineral, largely silica (silicon dioxide, SiO_2), which is similar to glass. One wrong blow and the hand axe shatters and is wasted. The people of Boxgrove knapped flint using harder stones; they also used soft hammers made of antler bone and probably used wooden tools too, but wood does not preserve well. All of this was long before people started using metals such as bronze.

The purpose of these hand axes has been hotly debated. At Boxgrove a local butcher was asked to cut up a carcass using a newly made axe. He found that the edge of the axe was sharp enough to give

Key summary: notch test – the energy needed to break a bone

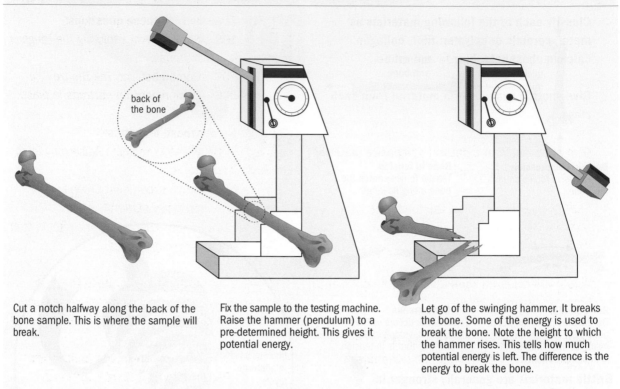

back of
the bone

Cut a notch halfway along the back of the bone sample. This is where the sample will break.	Fix the sample to the testing machine. Raise the hammer (pendulum) to a pre-determined height. This gives it potential energy.	Let go of the swinging hammer. It breaks the bone. Some of the energy is used to break the bone. Note the height to which the hammer rises. This tells how much potential energy is left. The difference is the energy to break the bone.

The toughness of a material is indicated by the energy needed to create the new fracture area

the clean cuts needed to do the job. When ancient axe heads were examined under a microscope, they were found to have wear markings just like those on the axe used by the butcher.

Sharp edges are characteristic of brittle, ceramic materials like flint. Ceramic materials, such as china, shatter under impact rather than deforming. At the same time, these materials can be very hard so they do not squash or dent. A bone hammer, on the other hand, is softer. It squashes slightly when struck. Its flexibility means that if you hit a bone hammer with great force it is more likely to bounce than break.

One of the lessons of the Boxgrove archaeological dig, and of many other digs at Stone Age sites, is that people have long used the materials they found around them to fashion useful artefacts. When the Boxgrove people killed a rhinoceros they looked around for suitable flints to make hand axes for their butchery work. Half a million years later we have ready access to artefacts made from a vast range of materials that come to us from all over the world.

Brittle bones

As we get older our bones undergo an ageing process too. They tend to become more porous and the proportion of calcium phosphate in them increases. Both of these factors make bones more brittle with age. That's why elderly people are particularly vulnerable to broken bones if they fall.

Brittle materials like glass and ageing bone fracture readily. Tests carried out on the femurs of people who died between the ages of three and 90 showed that the youngest bones absorbed three times as much energy as the oldest before they broke. Brittleness is caused by cracks spreading through a material. The opposite of brittleness is **toughness**. In young bones, collagen fibres help to stop cracks developing, so they are tough. In old bones, cracks travel easily through the brittle calcium phosphate, which is present in a larger amount than in younger bones. Engineers measure how easily a material fractures by finding the energy absorbed when a prepared specimen is broken. To measure toughness you need to know how difficult it is for cracks to deepen and spread.

Quick check

1. Classify each of the following materials as metal, ceramic or polymer: flint, collagen, calcium phosphate, bronze and rubber.

2. Give another example of a material from each class.

3. Give an example of a natural composite material and a synthetic composite.

4. Sketch diagrams to show the following:
   (a) the bricks of a house in compression;
   (b) a taut fishing line in tension.

5. Would you expect a shelf loaded with books to be in tension or compression?

6. Look around the room you are in now. Name two brittle materials and two tough materials.

Links to the *Advancing Physics* CD-ROM

 Practise with these questions:
10S Short answer *Exploring the range of materials*
20C Comprehension *The Bronze Age*
30C Comprehension *Portraits in plastic*

 Try out these activities:
90H Home experiment *Anisotropy in an apple*
100H Home experiment *Creep in an everyday material*
110H Home experiment *Time effects in an everyday material*

 Look up these key terms in the A–Z:
Ceramics; composite materials; glass; metals; polymers; stretching and breaking

 Go further for interest by looking at:
10T Text to read *Steel – the most important material?*
20T Text to read *Materials from nature*

 Revise using the revision checklist and:
150 OHTs *Materials selection charts*

4.2 Better buildings

Tree house

Timber is a traditional building material in most parts of the world. It is excellent stuff; it has the stiffness and strength needed to make a building two or more storeys high. This is not surprising; timber comes from trees, which themselves must be stiff and strong. A tall tree may be 30 m high and have a mass of 50 tonnes. It must be able to support its own weight, so the wood from which it is made must be strong in compression. At the same time, it must withstand high winds that try to bend it over, so it must also be strong in tension.

Wood is a natural composite material. It is made up of fibres of cellulose (the material of which the walls of plant cells are made), bound together by a substance called lignin. (Compare this with another natural composite, bone – described on p76.) Cellulose fibres are stiff and strong, and lignin prevents them from splitting apart.

Building high

In earlier decades, the height of skyscrapers was limited by their weight: the lower storeys could only just withstand the stress of the upper storeys pressing down on them. There were problems too with swaying in high winds. For a long time the Empire State Building in New York represented the limit of high-rise building. Now, developments both in materials (which are stiffer, stronger and more lightweight than before) and in design have resulted in a new generation of tall buildings.

Choosing building materials

Modern buildings can be made from a range of materials: stone, brick, steel, timber, glass and concrete. How can an architect set about deciding whether a particular material is suitable for the job? And what dimensions must a particular section have if it is to withstand the forces that will act on it?

Concrete is often used in the foundations of a building, and its strength in compression is about $50\,\text{N}\,\text{mm}^{-2}$. This means that each square millimetre of concrete can withstand a force of up to $50\,\text{N}$ pressing down on it – that's the same as the weight of 5000 tonnes pressing down on each square metre. A greater force would crush the concrete and a

A traditional house in Malaysia is made from timber. Suitable wood is readily available from local forests, and techniques have been developed over centuries for cutting and joining sections of timber to make houses with several rooms. It's relatively easy to add an extension, and so a house can eventually become a maze of interconnecting rooms.

The Petronas building in Kuala Lumpur, Malaysia, has 88 stories and stands 452 m high. The city-centre building represents Malaysia's claim to be a major player in world commerce.

The foundations of a building must be able to support the weight of the building above, so they must be strong in compression. Horizontal sections of the building's structure tend to bend under their own weight and because of the loads they have to bear. They are partly in compression and partly in tension, so materials must be chosen that are strong both in tension and in compression.

Tension and compression at home

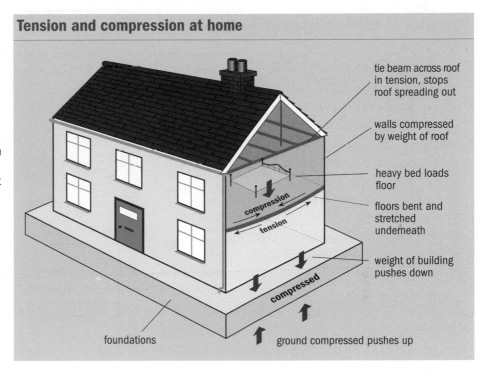

tie beam across roof in tension, stops roof spreading out

walls compressed by weight of roof

heavy bed loads floor

floors bent and stretched underneath

weight of building pushes down

compression

tension

compressed

foundations

ground compressed pushes up

building would be in danger of collapse.

Timber is a good material to use in horizontal sections (think of timber joists, which can be used to support floors), and so are metals such as steel. Concrete and brick are not as good to use here because they are much weaker in tension than they are in compression.

Metals have a different problem – they "give" before they break. As the load on a piece of steel is increased, it reaches a point at which the steel yields and becomes permanently deformed. This is the **elastic limit** or yield point, and an architect must ensure that the load on the steel is always less than this or it will bend and buckle. So it isn't the breaking stress (the amount of stress that breaks the metal) that determines a metal's useful strength; what matters is the stress at the yield point. **Breaking stress** and **yield stress** are two ways of thinking about the strength of a material.

Spiders' webs and steel

It's often said that the threads of a spider's web are stronger than steel – and it's true. This seems surprising, because you know how easy it is to brush aside a spider's web, while steel is known for its strength. If you compare a spider's thread with a steel thread of the same thickness, you find that the steel thread is weaker.

Key summary: force and stress

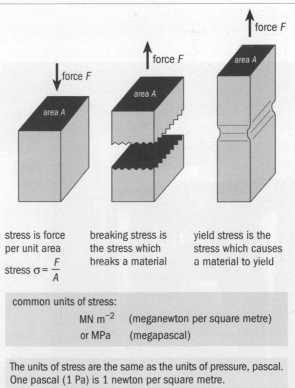

force *F*

force *F*

force *F*

area *A*

area *A*

area *A*

stress is force per unit area

stress $\sigma = \dfrac{F}{A}$

breaking stress is the stress which breaks a material

yield stress is the stress which causes a material to yield

common units of stress:

MN m^{-2} (meganewton per square metre)

or MPa   (megapascal)

The units of stress are the same as the units of pressure, pascal. One pascal (1 Pa) is 1 newton per square metre.

Useful rule of thumb: a mass of 1 kg weighs about 10 N on the Earth's surface.

Stress is force per unit area

Strengths of different types of materials

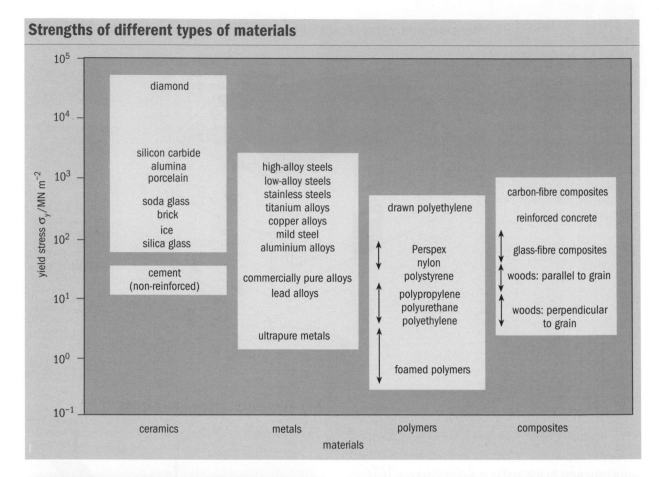

This building in Japan was designed to be strong. It has remained in one piece despite being toppled by an earthquake.

This piece of kit is being used to test a block of stone. It is controlled and monitored by a computer. Equipment like this is used to measure the strengths of materials. Metals such as steel are tested by stretching them in a tensile testing machine.

By comparing threads of the same thickness (or cross-sectional area), you are ensuring a fair test of the two materials. It's true that a spider's web is weaker than a steel bar, but the material that the web is made from is stronger than steel. It is important to distinguish between the strength of an object and the strength of the material from which it is made.

The box "Strengths of different types of materials" (above) shows the strengths of a number

Key summary: logarithmic scale of stress

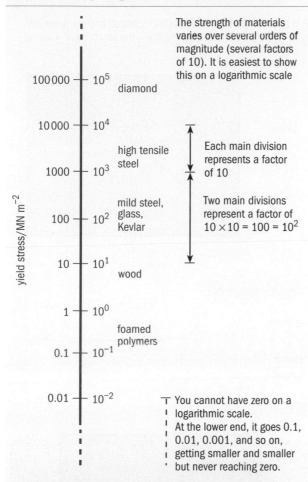

The strength of materials varies over several orders of magnitude (several factors of 10). It is easiest to show this on a logarithmic scale

Each main division represents a factor of 10

Two main divisions represent a factor of $10 \times 10 = 100 = 10^2$

You cannot have zero on a logarithmic scale.
At the lower end, it goes 0.1, 0.01, 0.001, and so on, getting smaller and smaller but never reaching zero.

Logarithmic scales are best used to display quantities that differ by several orders of magnitude

Key summary: stress produces strain

How much does a material stretch? For a given force F a long piece of material will stretch more than a short one. Doubling the length L doubles the extension x

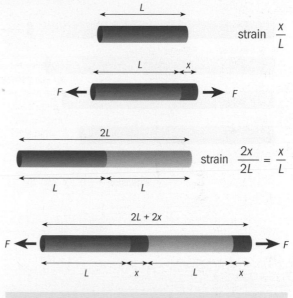

strain $\frac{x}{L}$

strain $\frac{2x}{2L} = \frac{x}{L}$

extension x depends on original length
strain = fractional increase in length and does not depend on the original length

$$\text{strain } \varepsilon = \frac{x}{L}$$

Units: strain compares lengths. If the units of extension and length are the same, strain ε is simply a number. It is often reported as a percentage, e.g. 1%

Tensile strain is a fractional increase in length

---

of different materials, grouped according to type. Things to note about this chart:

- the scale on the y axis is logarithmic, so each step represents an increase of a factor of 10;
- strength is represented as yield stress, σ_y in meganewtons per square metre $(\mathrm{MN\,m^{-2}})$;
- the strongest material (diamond) is at least 10^5 times as strong as the weakest (foamed polymers);
- cement is a weak material, but when it is made into concrete and reinforced by adding steel rods, it becomes much stronger.

Surviving an earthquake

During an earthquake, a building may shake back and forth so vigorously that it falls apart. The forces in it exceed the breaking strength of the

material from which it is constructed. We don't generally want to live in buildings that bend back and forth, but some flexibility is inevitable. In an earthquake zone, flexibility may even help by absorbing the energy of an earthquake and leave the building standing.

Building materials must be stiff – no-one wants to walk around on floors that stretch like trampolines. We don't want window panes that bow inwards when the wind blows, or outwards when we sneeze. But all materials have some degree of flexibility. The same testing machine that measures the strength of a material can be used to find out how flexible, or stretchy, it is; the result is known as the **Young modulus** of the material, though engineers often call it simply the elastic modulus. By knowing the Young modulus,

Key summary: the Young modulus

Many materials stretch in a uniform way. Increase the stretching force in equal steps, and the extension increases in equal steps too, in proportion. That is, the strain is proportional to the stress producing it. This is the same as Hooke's law – the stretching of a spring is proportional to the stretching force you apply.

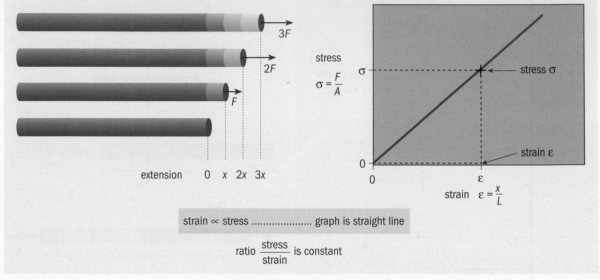

$$\text{strain} \propto \text{stress} \quad \text{.....................} \quad \text{graph is straight line}$$

$$\text{ratio } \frac{\text{stress}}{\text{strain}} \text{ is constant}$$

$$\text{Young modulus} = \frac{\text{stress}}{\text{strain}}$$

$$E = \frac{\sigma}{\varepsilon}$$

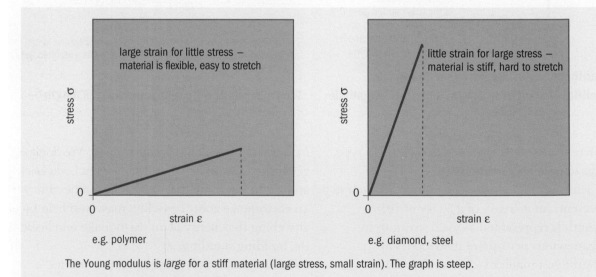

The Young modulus is *large* for a stiff material (large stress, small strain). The graph is steep.

The Young modulus is a property of the material not the specimen. Units of the Young modulus are MN m^{-2} or MPa, and for stiff materials GN m^{-2} or GPa, which are the same as units of stress, because strain is a ratio of two lengths, e.g. extension is 1% of length.

The Young modulus is a stress divided by strain ratio

designers can calculate how thick a beam, or other building component, must be to support the weight of any load it is likely to have to bear, without bending too much.

Rubber and jelly are examples of materials at the other end of the scale of stiffness. They are flexible, and bend or stretch very easily. Plastics like polythene are a bit stiffer.

Beyond the elastic limit: testing window glass at the British standards Institute. The wooden testing device shown represents a human head.

Specially shaped metal samples are put in this tensile testing machine and they are stretched gradually. The strain recorded increases at a steady rate.

Stress versus strain

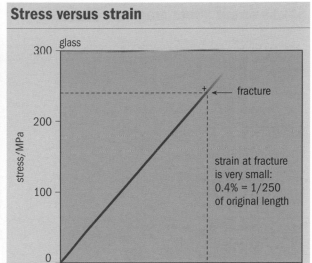

A stress versus strain graph for glass is a straight line up to the point where the sample fractures. At breaking point the strain is less than 1% – the sample's length increases by this tiny amount.

Breaking glass, stretching steel

It can be hard to believe that glass is a flexible material. If you press gently on the middle of a window pane, it will bow outwards slightly. The glass fibres of a surgeon's endoscope can bend, following the curves of a patient's insides.

A small stress will cause glass to stretch slightly. Remove the stress and the glass returns to its original dimensions. This is **elastic deformation**. Increase the stress too much and the glass breaks. This is **brittle fracture**.

Some metals are brittle – they behave like glass. Others, such as mild steel, are different. Think of what happens when a car crashes. The steel bodywork is easily dented, but it is unlikely to break.

For very small strains, mild steel shows elastic behaviour, like a spring. But then, at the elastic limit, it yields. Now it begins to deform – this is **plastic deformation**. The strain may increase to 40% or more before the metal eventually fractures.

Wood ready to wear

This section started by discussing wood for buildings, and now ends with wood as a source of fibres. A family of four would be able to obtain a lifetime's worth of clothing from just one tree. However, for this to work you would really have to like viscose – the material that is manufactured

Stress versus strain graph for mild steel

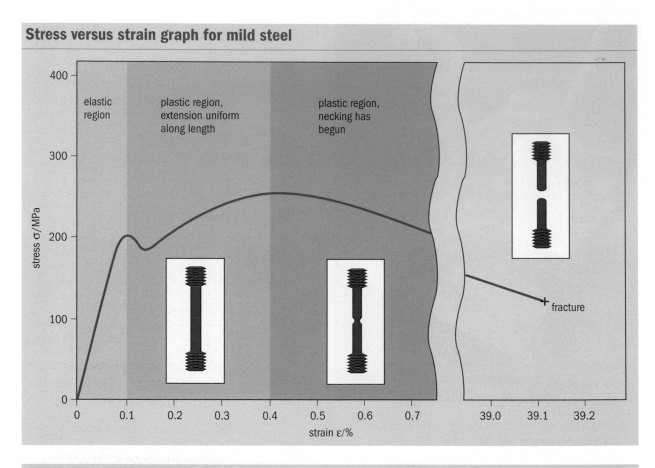

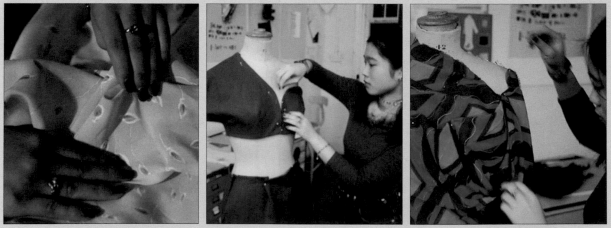

Almost all clothing is made from polymers. We rely on the flexibility of material for a figure-hugging fit. Other materials must be stiff for things like the sole of a shoe. Even the natural fabrics we wear — such as cotton and leather — are polymers.

from the bark of eucalyptus, pine and beech trees. Viscose and its close relations acetate and triacetate are all thin, soft, slightly shiny fabrics that are often used to make dresses and to line suits. But can you imagine having all your clothes – underwear, socks, trousers, jackets and skirts – made from just this one material?

Viscose is a polymer. The fact that viscose clothing will stretch to fit around your body shows that it has a degree of elasticity. In fact, polymers tend to be more elastic than most metals or ceramics. But don't be misled by the way we often refer to some polymers as plastics. Although some polymers show plastic behaviour, others are brittle. And their behaviour can change with temperature. At room temperature, Perspex is brittle like glass. Warm it up and it becomes bendy. Warm it some more and you can deform it to any shape you like.

Quick check

The first four questions relate to the chart on p82.

1. Which class of materials is generally the strongest?

2. Which are stronger, pure metals or alloys?

3. "Ice" is a slang name for diamonds. Show that diamond is 1000 times stronger than ice.

4. Show that reinforced concrete is 10 times stronger than cement that has not been reinforced.

5. A concrete block of cross-sectional area $0.01\,m^2$ is crushed when a force of $1\,MN$ is applied to it. Show that its breaking stress is
   (a) $100\,MN\,m^{-2}$;
   (b) $100\,N\,mm^{-2}$.
   Recalling that $1\,kg$ weighs about $10\,N$, also show that the block can support up to $100\,000\,kg$.

6. Using the stress versus strain graph for glass on p85 show that
   (a) the fracture stress is about $240\,MPa$;
   (b) the Young modulus is about $60\,GPa$.

7. Using the stress versus strain graph for mild steel on p86 show that
   (a) the fracture stress is about $125\,MPa$;
   (b) the Young modulus is about $200\,GPa$.

Links to the *Advancing Physics* CD-ROM

Practise with these questions:
10E Estimate *Making estimates about the mechanical behaviour of materials*
45S Short answer *Calculations on stress, strain and the Young modulus*
50S Short answer *Measuring the Young modulus*
50D Data handling *Stress, strain and the Young modulus*

Try out these activities:
100E Experiment *Plot and look: measuring breaking stress of materials*
150E Experiment *Good measurements of stiffness and strength (Young modulus and breaking stress) of materials*

Look up these key terms in the A–Z:
Mechanical characteristics of materials; solid materials; stretching and breaking

Go further for interest by looking at:
50T Text to read *Fantastic fibres*

Revise using the revision checklist and:
200 OHT *Forces of tension and compression*
400 OHT *Strengths of some materials*
500 OHT *The Young modulus*
600 OHT *Stress–strain graph for mild steel*

4.3 Conducting well, conducting badly

Some materials are better electrical conductors than others. The scale of conductivity below shows that some materials are excellent conductors of electricity. The best conductor (silver) has a conductivity that is almost a million million million million (10^{24}) times as good as that of the best insulator (polystyrene).

Electrical conductivity varies over a wider range than any other material property. Because of the great range of values, the chart in the Key summary below uses a logarithmic scale.

Key summary: logarithmic scale of resistivity and conductivity

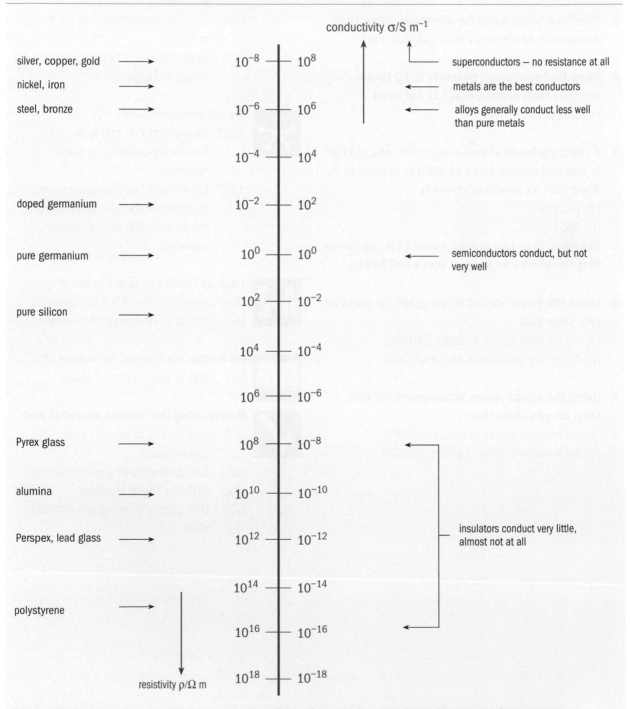

A logarithmic scale is essential to display quantities that vary by more than 20 orders of magnitude

Key summary: conductivity and resistivity

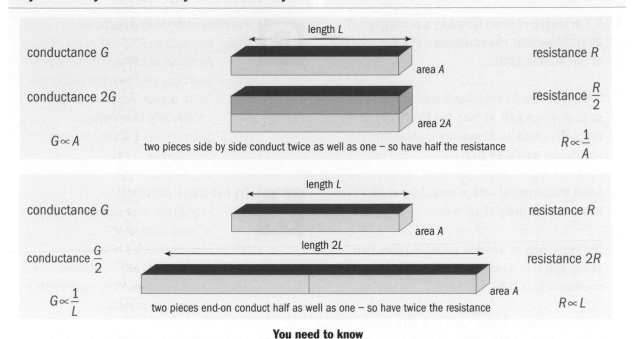

conductance G length L resistance R

area A

conductance $2G$ resistance $\dfrac{R}{2}$

area $2A$

$G \propto A$

two pieces side by side conduct twice as well as one – so have half the resistance

$R \propto \dfrac{1}{A}$

conductance G length L resistance R

area A

conductance $\dfrac{G}{2}$ length $2L$ resistance $2R$

area A

$G \propto \dfrac{1}{L}$

two pieces end-on conduct half as well as one – so have twice the resistance

$R \propto L$

You need to know
length L
cross-sectional area A

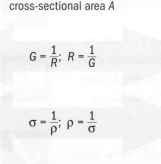

to find
conductance G from conductivity σ

$G = \dfrac{\sigma A}{L}$

unit: siemens (S)

$G = \dfrac{1}{R}$; $R = \dfrac{1}{G}$

to find
resistance R from resistivity ρ

$R = \dfrac{\rho L}{A}$

unit: ohm (Ω)

conductivity σ from conductance G

$\sigma = \dfrac{GL}{A}$

unit: S m^{-1}

$\sigma = \dfrac{1}{\rho}$; $\rho = \dfrac{1}{\sigma}$

resistivity ρ from resistance R

$\rho = \dfrac{RA}{L}$

unit: Ω m

Conductivity σ and resistivity ρ give the same information in complementary ways

Pairs of quantities

Artefact property	Unit	Material property	Unit
mass (of an object)	kg	density	kg m^{-3}
conductance (of a wire)	S	conductivity	S m^{-1}
resistance (of a wire)	Ω	resistivity	Ω m
stiffness (of a spring)	N m^{-1}	Young modulus	N m^{-2}
force to permanently deform (a metal bar)	N	yield stress	N m^{-2}
force to break (a window)	N	breaking stress	N m^{-2}

Materials and artefacts

Which is heavier, a tonne of lead or a tonne of feathers? Of course, both have the same mass but lead is more dense. In this chapter you have seen examples of how to distinguish between the properties of an individual object and the properties of the material from which it is made. Mass and density are two properties where care is needed. The mass of an object depends on its dimensions as well as the material it is made from.

Quick check

1. A 1 m length of steel wire has a resistance of 20 Ω. Show that the resistance of a 5 m length of this wire is 100 Ω.

2. If the 1 m wire in question 1 was folded in half, so that it was half as long but in effect has twice the cross-sectional area, show that its resistance would be 5 Ω.

3. Show that a metal with a resistivity of 10^{-8} Ω m has a conductivity of 10^8 S m^{-1}.

4. The resistivity of another metal is twice that of the metal in question 3. Show that its conductivity is 0.5×10^8 S m^{-1}.

5. A polymer fibre, 100 m long and with cross-sectional area 10^{-6} m^2, has a resistance of 10^{18} Ω. Show that its resistivity is 10^{10} Ω m.

6. Use information in the Key summary on p88 to show that you could expect the resistance of a sample of doped semiconductor, length 10 mm, width 5 mm and thickness 1 mm, to be about 200 Ω.

7. The resistivity of copper is about 10^{-8} W m. Show that a 100 km-long copper power line 10 mm in diameter has resistance of the order of 10 Ω.

Links to the *Advancing Physics* CD-ROM

Practise with these questions:

20E   Estimate *Making estimates about electrical behaviour of materials*

70S   Short answer *Electrical properties*

80S   Short answer *Resistivity and conductivity calculations*

100S   Short answer *Conductance and conductivity*

Try out these activities:

310E   Experiment *Measuring resistance of good conductors*

330E   Experiment *Measuring the resistance of two insulators*

350E   Experiment *Good measurements of electrical resistivity*

Look up these key terms in the A–Z:

Conductance; electrical conductivity and resistivity; resistance

Go further for interest by looking at:

100T   Text to read *Introduction to materials selection charts*

Revise using the revision checklist and:

100O   OHT *Range of values of conductivity*

110O   OHT *Conductivity and resistivity*

4.4 Problems of measuring mechanical and electrical properties

This section looks at some of the problems of measuring:

• breaking stress;

• the Young modulus;

• and electrical resistivity or conductivity.

These problems illustrate difficulties found in many other kinds of measurement. See Case studies: quality of measurement (p219) for a variety of other examples.

Problems of measuring breaking stress

Cotton thread can certainly break under stress, as you find out when a button catches on something and is pulled off your coat. The chart below shows the variation in results obtained by a class measuring the force needed to break each of 30 samples of the same cotton thread. The threads were pulled on with a newtonmeter, gradually increasing the tension until each thread broke.

Faced with this kind of variation between results, as illustrated in the chart, you have to ask why they vary. Here, the reason is not uncertainty in measuring the forces: the newtonmeters used could each record accurately to better than ±0.5 N. This is an example of natural, or inherent, variation

between samples. The various sample threads had differing breaking forces.

If you think about it, it is easy to understand why samples of cotton thread could break at different forces. A break will usually start at a weak point in the material, for example where it is thinner than usual, or where some fibres are broken. For this reason, breaking stress is not a well defined property. It may be more important to know the range over which the breaking stress varies than the exact value of the average for many samples. This would give a safe value if you were specifying the thickness needed for a given application.

There is a further problem. Stress is equal to the force divided by the cross-sectional area. Because these threads are very thin, in fact little more than 0.1 mm, it is difficult to measure their diameter accurately. Even using a micrometer the uncertainty could easily be ±10%. Worse, the uncertainty in the cross-sectional area is twice

Key summary: change in diameter and area

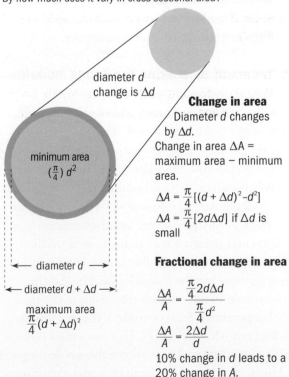

A thread varies in diameter.
By how much does it vary in cross-sectional area?

diameter d
change is Δd

minimum area
$(\frac{\pi}{4}) d^2$

←— diameter d —→

←— diameter $d + \Delta d$ —→

maximum area
$\frac{\pi}{4}(d + \Delta d)^2$

Change in area
Diameter d changes by Δd.
Change in area ΔA = maximum area − minimum area.

$\Delta A = \frac{\pi}{4}[(d + \Delta d)^2 - d^2]$

$\Delta A = \frac{\pi}{4}[2d\Delta d]$ if Δd is small

Fractional change in area

$\frac{\Delta A}{A} = \frac{\frac{\pi}{4}2d\Delta d}{\frac{\pi}{4}d^2}$

$\frac{\Delta A}{A} = \frac{2\Delta d}{d}$

10% change in d leads to a 20% change in A.

The fractional uncertainty in a quantity is doubled if the quantity is squared

Breaking cotton

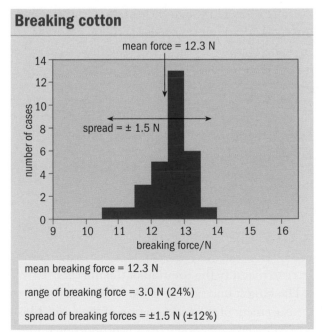

mean force = 12.3 N

spread = ± 1.5 N

number of cases

breaking force/N

mean breaking force = 12.3 N

range of breaking force = 3.0 N (24%)

spread of breaking forces = ±1.5 N (±12%)

Some 30 samples of thin cotton thread from the same reel were each pulled until they broke. The force needed to break each sample was recorded.

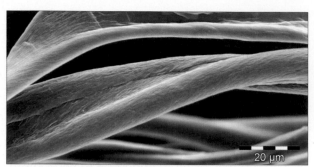

Fibres of cotton seen in a scanning electron microscope. Cotton threads are made of these fibres twisted together. A thread does not have a sharply defined diameter.

as large as this, say $\pm 20\%$, because the area is proportional to the square of the diameter.

This last conclusion can be generalised. If a quantity appears to the power n in the calculation of a result, its percentage of uncertainty is multiplied by n when calculating percentage uncertainty of the result.

On top of this, if you look at a cotton thread under a microscope you see that there simply is no exact diameter at all. The surface of the thread is fuzzy, with no sharp well defined edge.

This is a case where measuring as well as possible does not mean trying to be more and more precise. All you can do is to measure to within the natural variations that exist.

Problems of measuring the Young modulus

The main problem in measuring the Young modulus is that most materials stretch very little, even under quite large forces. They often yield at a strain of only a few per cent, or less. And a good thing too, because this allows us to build strong, stable structures, such as houses, cranes and towers. But it means that to measure the Young modulus you have to measure a very small extension.

You can't design a way to measure a small extension as well as possible unless you know roughly how big the extension will be. So the golden rule in any measurement has to be: do a trial experiment first, or make a rough estimate, to find out what to expect. You can't plan to hit a bulls-eye until you know where the whole target is.

For steel, a rough estimate or a trial experiment quickly tells you two things:

- for any reasonable length of wire you will have to measure an extension of only a few millimetres;

Key summary: estimating a Young modulus

Approximate values for steel:
yield stress 1 GPa
the Young modulus 100 GPa

Estimate of extension
The yield stress is 1% of the Young modulus, so the strain at the yielding point is about 1% and the extension to be measured is <1% of the length. If the length is 1 m, the extension is <10 mm.

Estimate of force needed
At 1% strain, the stress is about $10^9\,\mathrm{N\,m^{-2}}$. This is very large, so try a $1\,\mathrm{mm^2}$ cross section $= 10^{-6}\,\mathrm{m^2}$:

To achieve this with reasonable force (say 100 N) you need a cross section of about $0.1\,\mathrm{mm^2}$. So try wire of about 0.3 mm diameter.

Start with a rough estimate or experiment to find out what to expect

- to stretch the wire with any reasonable force you will have to use a thin wire, a fraction of a millimetre in diameter.

So, to measure the Young modulus as well as possible you will have to measure two rather small distances: the extension and the diameter of the wire. Both being small, they are both hard to measure with small uncertainty.

Your laboratory may have devices such as vernier scales, micrometers and travelling microscopes for resolving small distances. You need to check them to see if they are up to the job and to estimate the uncertainty they will leave in the measurement of extension and diameter.

It may be possible to be creative in finding ways to do better. How about making the stretching wire turn a pulley with a mirror attached to the pulley, and using a long beam of light as an optical lever to magnify the rotation of the mirror?

The largest uncertainty is likely to be in the measurement of the cross section of the wire. This is because cross-sectional area is proportional to the square of the wire diameter, so the percentage uncertainty in the cross-sectional area is double

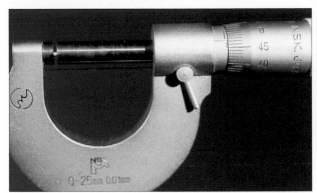

A micrometer screw gauge for measuring small thicknesses.

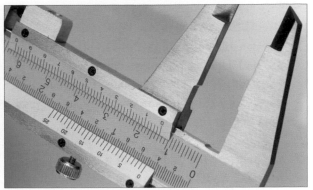

Vernier calipers for measuring small spacings or thicknesses.

the percentage uncertainty in the diameter (see the Key summary on p91). So anything you can do to reduce this uncertainty will be repaid. For example, if you knew the density of the steel, you could weigh a long length of it with less uncertainty and find the cross section that way.

It is easy to make a systematic error in measuring the extension of a wire under tension. Because the force needed is quite large, the supports of the fixed end of the wire can easily give a little, making the extension seem larger than it really is.

Problems of measuring resistivity

Always make a rough measurement or estimate to locate the target. Values of resistivity or conductivity vary over at least 20 orders of magnitude (p88) from good conductors like metals to good insulators like perspex. So the problems you face will be very different at each end of this range.

For insulators, the problem will be in detecting any electric current at all. If you do get a current, it is quite possible for it to be due to conduction through moisture or contamination on the surface of the material rather than due to conduction through the material. You will need a short fat specimen – a slab rather than a wire – and it will have to be kept dry and clean. You will also need a sensitive ammeter. In fact, it may not be possible to measure the conductivity of such a material at all, other than to show that the conductivity must be smaller than the smallest value you can measure.

This is not necessarily a waste of time. Many measurements at the cutting edge of physics turn out this way.

For good conductors the problem is getting a

specimen long enough and thin enough to have an appreciable resistance. For example, the resistivity of copper is about $10^{-8}\,\Omega\,\mathrm{m}$. Thus a 1 m length of wire of cross section $1\,\mathrm{mm}^2$ can be expected to have a resistance of only $0.01\,\Omega$.

$$\text{Resistivity } \rho \approx 10^{-8}\,\Omega\,\mathrm{m}$$
$$\text{Length } l \approx 1\,\mathrm{m}$$
$$\text{Cross section } A \approx 1\,\mathrm{mm}^2 = 10^{-6}\,\mathrm{m}^2$$

$$\text{Resistance } R = \frac{\rho l}{A} \approx \frac{10^{-8}\,\Omega\,\mathrm{m} \times 1\,\mathrm{m}}{10^{-6}\,\mathrm{m}^2}$$

$$= 10^{-2}\,\Omega$$

Clearly, you will need a wire thinner and perhaps longer than this to reach a resistance of a few ohms, which should be possible to measure easily. Now, however, the main uncertainty will be in the measurement of the wire diameter, and so the cross sectional area.

You may need to think about how best to measure the resistance of the wire. Instead of an ammeter and voltmeter combination, a digital ohm-meter might be better, but only if you know that you can rely on it. That means testing it on a resistor of known resistance. Or, if you have such a resistor, you might place it and the wire in a potential divider and measure the p.d. across each. Which method is best depends on what instruments you have and on whether you have standardised resistors available.

Systematic error is common in electrical measurements. For example, the electrical contacts to the wire themselves have some resistance, which may make the resistance of the wire seem larger than it really is.

Quick check

1. The densities of a number of samples of timber are (in kilograms per metre cubed – $kg\,m^{-3}$): 750; 700; 740; 740; 760. Show that the mean density is $738\,kg\,m^{-3}$ and that the range is less than 10% of the mean.

2. You measure the thickness of a sheet of paper to be $0.2 \pm 0.04\,mm$. Show that the percentage uncertainty is $\pm 20\%$.

3. You measure the thickness of square cross section of rubber cord using a ruler. You get the result $5.0 \pm 0.5\,mm$. Show that the percentage uncertainty in the cross-sectional area is $\pm 20\%$.

4. In measuring the Young modulus of copper wire you obtain these values: length of wire = $1.5 \pm 0.005\,m$; diameter of wire = $0.33 \pm 0.02\,mm$; extension = $3.5 \pm 0.02\,mm$; tensile force = $11.5 \pm 0.2\,N$. Show that the largest percentage uncertainty is $\pm 6\%$. Explain why this uncertainty is particularly important in this case.

5. You are measuring the resistivity of a thin layer of graphite deposited on a strip of paper $102 \pm 1\,mm$ long, $9.0 \pm 1\,mm$ wide. You estimate the thickness of the graphite layer to be somewhere between 0.1 and 0.2 mm. The resistance of the strip along its length is $3.5 \pm 0.1\,\Omega$. Show that the largest percentage uncertainty among these values is about $\pm 30\%$.

6. In the experiment of question 5, electrical contact has to be made with the graphite. Explain why this introduces a systematic error.

Links to the *Advancing Physics* CD-ROM

Practise with this question in *Case Studies: quality of measurement*:
210D Data handling *Calculating with uncertainties: chapters 4–5*

Try out these activities:
100E Experiment *Plot and look: measuring breaking stress of materials*
150E Experiment *Good measurement of Young modulus*
350E Experiment *Good measurements of electrical resistivity*

Look up these key terms in the A–Z:
Average; systematic error; uncertainty

Go further for interest by looking at:
Case studies: quality of measurement (p219)

Revise using the revision checklist

Summary check-up

Classes of materials ✓

- Classes of materials include metals, ceramics and polymers
- Composites are designed to combine desirable properties from two or more materials

Elastic materials ✓

- Solid materials behave elastically up to the elastic limit
- The Young modulus is the gradient of the initial, linear part of the stress versus strain graph
- Young modulus = $\frac{\text{stress}}{\text{strain}}$

  where stress = $\frac{\text{load}}{\text{area}}$,

  strain = $\frac{\text{increase in length}}{\text{initial length}}$

How materials fail ✓

- Metals deform plastically before they break. A lot of energy is needed to break them – they are tough materials
- Brittle materials (notably ceramics, including glass) break easily, without deforming plastically
- Some polymers are brittle; some are very elastic; some are very tough

Electrical behaviour ✓

- Electrical conductivity tells you how well a material conducts an electric current
- Electrical conductivity varies widely between materials
- Electrical resistivity ρ is the reciprocal of conductivity σ:

$$\rho = \frac{1}{\sigma}$$

- To calculate the resistance R of an object knowing the resistivity ρ of the material from which it is made, use

$$R = \frac{\rho L}{A}$$

- To calculate the conductance G of an object knowing the conductivity σ of the material from which it is made use

$$G = \frac{\sigma A}{L}$$

Measurement of properties ✓

- A property may vary from one sample to another
- To improve a measurement, think about the value with the greatest uncertainty
- Be aware of the possibility of systematic error

Questions

1. **Most cups or mugs used at home are made from china; picnic cups are usually plastic.**
   (a) Explain why these different materials are used; refer to these properties: density, stiffness, brittleness, and any other properties that you consider important.
   (b) What key properties would be important for a material used for dental fillings? Name two materials commonly used for this purpose.

2. **Look at the five objects A–E. Each is being pulled by the force shown. Explain why B will be stretched more than A, C more than B, D more than C, and E more than D.**

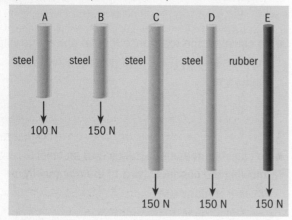

3. **A rod is made from a type of glass whose breaking stress is 40 MPa. The cross-sectional area of the rod is 100 mm^2 (10^{-4} m^2).**
   (a) What is meant by the term breaking stress?
   (b) What force is needed to break the rod?

4. **Steel has a conductivity of 10^6 S m^{-1}, a Young modulus of 200 GPa and a yield stress of 200 MPa.**
   (a) Estimate the radius and cross section of an unstretched 1 m long steel wire with resistance 1/3 Ω.
   (b) Estimate the tension in this wire when it yields.
   (c) Assuming that the extension is linear up to the yield point, estimate by how much the wire has stretched just before it starts to yield.
   (d) Why does the resistance of the wire change as it is stretched? Will the change be large?

5. **A rubber cord, 5 m long and 50 mm thick, hangs from the top of a zoo cage. A chimp weighing 300 N hangs on to the end of the cord. How long is the cord now? (The Young modulus of the rubber is 100 MPa.)**

6. **In measuring the Young modulus of a rubber band in its initial region of linear strain the following measurements were taken: force $F = 0.5$ N ± 1%; cross-sectional area $A = 4.0$ mm^2 ± 3%; extension $x = 0.0080 ± 0.0005$ m; length $L = 0.145 ± 0.001$ m. Calculate the Young modulus and make some estimate of its percentage uncertainty.**

7. **A digital ohm-meter connected to a resistor displays three digits. The last digit flickers, a series of 10 readings from it being (in ohms): 97.4, 97.5, 97.1, 97.9, 97.3, 97.0, 97.5, 97.4, 97.3, 97.6. State the percentage uncertainty that you would attribute to the mean value. What would be the smallest systematic error in resistance that you would need to consider trying to eliminate?**

8. **The table below shows the resistance of 1 m lengths of various wires, each 0.25 mm in diameter and made from different materials.**

material	resistance/Ω
copper	0.351
Eureka	10.0
manganin	8.45
nichrome	22.0

   (a) Which of these materials has the greatest resistivity?
   (b) Calculate the resistivity of this material.

9. **Use the chart of conductivity on p88 to find how much better a conductor copper is than steel. How do semiconductors get their name?**

5 Looking inside materials

Damascus steel for swords and Chinese porcelain for ornaments are examples of how properties of materials have been improved by trial and error. Today, a better understanding of how the structure of materials decides their properties leads to designer materials, with purpose-built structures providing desired properties. Old needs are better met, and new uses opened up. In this chapter we will give examples of:

● how the structure of a material determines its properties
● and how materials can be designed to have desirable properties

5.1 Materials under the microscope

How can you find out about the structures of materials? Look at a piece of wood. Its grain reveals its structure. Wood for furniture is usually cut so that the grain runs straight down the legs of a chair, or along the length of a table top. Take a wooden ice-lolly stick and split it end to end. The wood breaks easily between the fibres that make up the grain. Break the stick across its length and you will be left with rough ends, with stiff fibres exposed. Under an optical microscope, the long, narrow cells that make up the wood are revealed. It is these cells that give wood its characteristic structure.

You can see the structure of wood. Carpenters pay attention to it and avoid pieces of wood with knots or distorted grain as these would form areas of weakness in a chair leg, for example.

But what about materials where the structure is not obvious? The metal of a coin or the plastic of a Lego brick appear entirely uniform to the naked eye. Today there are several different types of microscope that can help to show the internal structures of such materials.

The pictures on p98 are examples of images obtained from different kinds of microscopes. They reveal different aspects of material structures.

A snow crystal magnified 6500 times in a scanning electron microscope (SEM), showing the intricate branching shape of the ice crystals in snow.

A question of materials

As physicists dissected the anatomy of atoms... all kinds of engineers and scientists delved in ever more detail into the relationship between a material's many-levelled structure and its properties. Why exactly do ceramic materials break so readily but metals only dent? Why does steel resist rusting when you put enough chromium into it? What is going on in a piece of metal when it undergoes age-hardening? What makes rubber so elastic? Can a carefully constructed crystalline material function like an entire device, say, like the vacuum tubes of early electronic technology? ...Like a person's personality, every material's collection of traits and behaviours comes from what it is made of and how those ingredients were nurtured into their present form.

Ivan Amato 1997 Stuff: The Materials the World is Made of *(Avon Books)*

Order, order

Atomic force microscopes (AFMs) and scanning tunnelling microscopes (STMs) allow us to "see" the atoms of which a material is made. They are called microscopes, but they don't allow us to look directly at a sample like an optical microscope. Instead, they scan the surface of the sample, building up an image on the screen of a monitor.

These microscopes tell us about the arrangement of atoms on the surface of a material. But

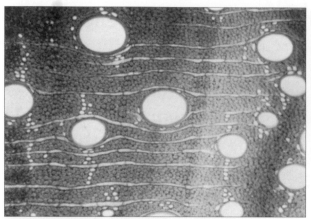

Viewed under an optical (light) microscope, wood is shown to be made of many long, parallel cells that give the material its characteristic grain structure.

In this unusual image an ant is seen by a scanning electron microscope (SEM) carrying a tiny electronic microchip. An advantage of an SEM over an optical microscope is that the whole insect can be in focus. A disadvantage is that the image can only be seen in black and white.

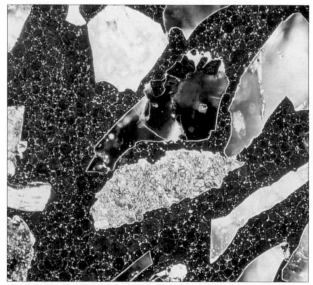

An optical microscope was used to obtain this colourful image of a "cermet" – a composite material. You can see that angular ceramic particles are embedded in steel.

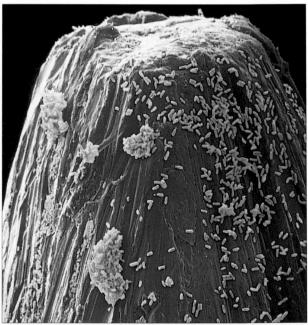

This is a false-colour SEM image of bacteria (yellow) on the point of a pin. It has been coloured to show the different features. Notice how the steel of the pin, which would feel smooth to the touch, is in fact rough when seen on this scale.

The atoms in a metal usually form a regular, crystalline array as shown in this atomic force microscope (AFM) image. The microscope can only just resolve the atoms. The conical appearance is an artefact of the imaging process.

A scanning tunnelling microscope (STM – p8) allows us to see matter on the scale of individual atoms. Here, graphite atoms (green) form a regular array, with atoms of gold (orange) piled up on the graphite surface. This is a false-colour image.

Key summary: how a scanning electron microscope (SEM) works

very narrow
electron beam
scans specimen

narrow
electron
beam

scan lines

amplifier and monitor

monitor scanned
in time with
beam scanning
specimen

collector of
electrons

specimen

bright spot on
monitor screen
corresponds to
spot on specimen

$$\text{magnification} = \frac{\text{scan line spacing on monitor}}{\text{scan line spacing on specimen}}$$

Pros and cons The specimen must be in a vacuum and be coated with a metal film, so the specimen cannot be alive. The SEM has a good depth of focus and shows relief very well.

Scanning electron microscopes can provide stunning three-dimensional images

A fine beam of electrons scans a tiny specimen, which is usually coated with metal. From the place where the beam is on the specimen, electrons are scattered or emitted and picked up by a collector. The signal from the collector controls the brightness of a spot scanning a monitor in time with the beam scanning the specimen. A picture of the specimen, point by point, is built up on the monitor screen. The magnification just depends on how much bigger the monitor screen is than the scanned area of the specimen.

beware, the atoms inside the material may be arranged differently. Surface atoms often rearrange themselves because, unlike the atoms inside a material, they are not surrounded on all sides by other atoms. Techniques such as X-ray crystallography may tell us more about the arrangement of atoms inside a material.

All these methods can show us the regular, orderly arrangement of the atoms that make up a metal. Metals are crystalline, though they usually don't look like crystals.

In fact, for centuries people have supposed that the atoms in a metal are arranged in a regular, crystalline array. There's a story about this. Robert Boyle (of Boyle's law fame) was on the quayside, watching the loading of warships. He saw neat stacks of cannonballs and noted how their

arrangement gave rise to planes and angles just like those he had seen on the surfaces of crystals. Could crystals be made from millions of tiny, hard spherical particles?

This story may be apocryphal, but it does illustrate how people can take an observation and use it to jump to a new idea in some apparently unrelated area. Today, three centuries later, there are new ways of looking at atoms and seeing their arrangements that are, for many materials, just as Boyle suggested.

Measurements show that the smallest atoms measure about 0.1 nm in diameter and the largest atoms are about 0.5 nm ($1\,\text{nm} = 10^{-9}\,\text{m}$). So, for the largest atoms, about 2 000 000 will fit side by side in just one millimetre.

Try this: put some marbles in a shallow dish

Key summary: how an atomic force microscope (AFM) works

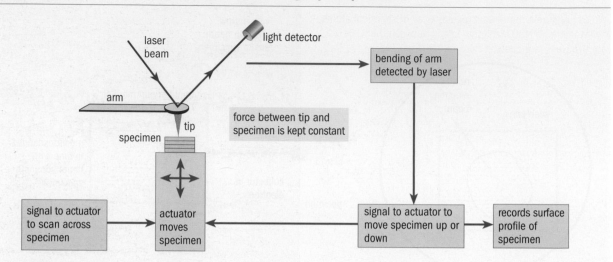

Pros and cons The AFM does not need the specimen's surface to be specially prepared. The specimen doesn't need to conduct electricity or be in a vacuum. The AFM is suitable for biological work.

Atomic force microscopes can be used to analyse a variety of specimens

Much like an old-fashioned record player, an atomic force microscope moves a needle over a sample to detect the contours of the surface. In this case surface roughness is detected on an atomic scale. A fine point is mounted on a cantilever arm (see bottom image on p98). Forces between tip and surface bend the cantilever. A laser beam reflected from the cantilever detects the bending. In one mode of operation the specimen is moved to keep the force on the tip constant. Up and down movements of the specimen as it is scanned under the tip correspond to the surface profile.

or saucer. (If you don't have marbles, try lentils or dried peas.) Look at the regular pattern they make. Now try grains of rice, short grain and long grain. What sorts of patterns do you see? Marbles are like the spherical atoms of which metals are made; rice grains are more like the elongated molecules of a liquid crystal. Polymers are made of flexible, long-chain molecules. How could you model polymers in the kitchen?

In the box "Models of metals" (opposite) the two images show some realistic features of the arrangements of atoms in metals. But, like any model, the comparison has its limitations. There is no attractive force between the ball bearings: tip the dish and they roll apart. And both models are simply two-dimensional representations of a three-dimensional reality. Modern microscopy allows us to see images of atoms.

Crystals you can see and crystals you can't

Solids form when a liquid cools. The internal structure of the resulting solid may be amorphous (disordered) or crystalline. Rapid cooling tends

to trap particles in an amorphous state, which resembles a disordered arrangement in a liquid. Glass is an example of an amorphous material. On the other hand, slow, controlled cooling can produce a single crystal, which is a highly ordered array of atoms. An important example of a single-crystal material is high-purity silicon, which is used for making microchips. A single crystal may weigh several kilograms and be made up of 10^{27} atoms arranged in an almost perfect array.

Many materials, however, are neither completely amorphous nor perfectly crystalline in the solid state; rather, they are polycrystalline. As a liquid cools, crystals start to form at different points within it. Each crystal grows out into the remaining liquid until it runs into its neighbours. The result is a patchwork of tiny crystals or grains. Where these grains meet, the interface is known as a grain boundary. Therefore a polycrystalline material consists of a number of grains all orientated differently relative to one another but with an ordered structure within each individual grain.

Models of metals

We can picture a metal as a regular array of spherical atoms held together by the attractive forces between them. Here are two ways of modelling this in the lab.

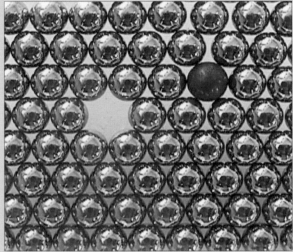

Ball bearings in a dish This model shows defects in the lattice: vacancies where the atom is missing and interstitials where a larger atom has been inserted.

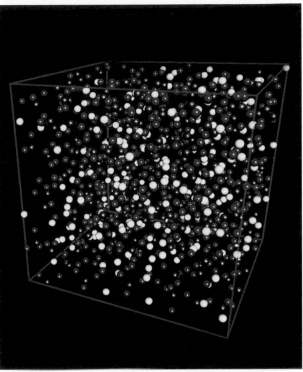

This model shows how a typical arrangement of atoms in sodium disilicate glass. This is an amorphous structure. The positions of the atoms were calculated by computer so that they reproduce the results of detailed X-ray studies of the material.

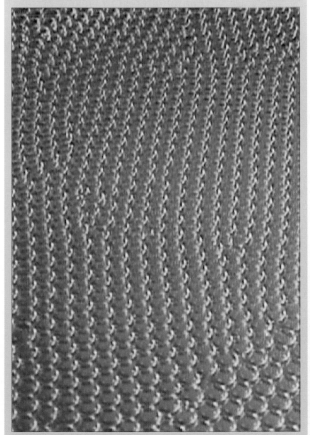

Raft of soap bubbles Burst a bubble and this model will show a vacancy. It also shows dislocations well. You can even watch as planes of bubbles slide past one another.

The polycrystalline structure of a sample of brass can only be seen using a light microscope.

Quick check

1. Give an example of an amorphous and a polycrystalline material.

2. An atom of iron is about 0.25 nm in diameter. Show that 1 million will fit across the 0.25 mm diameter of a pin head.

3. How many nearest neighbours does an "atom" have in the ball bearing model shown on p101? Is the answer the same for the soap-bubble model on the same page?

4. In a simple cubic arrangement the atoms sit at the corners of imaginary cubes with their curved surfaces just touching. Show that the ratio of filled space to empty space between the atoms is $\frac{\pi}{6}$ or about 0.5. This "packing fraction" is related to the density of the material. The volume of a sphere of radius r is $\frac{4}{3}\pi r^3$.

5. If the atoms in the simple cubic arrangement described in question 4 have a diameter d, show that the largest impurity atom that can just fit in between the host atoms has a diameter of $0.4d$.

6. The monitor screen of a scanning electron microscope is 200 mm × 200 mm, and is scanned in 500 lines. If the corresponding area of a specimen is 1 μm × 1 μm, show that the linear magnification is 100 000. Also show that the system can just resolve 2 nm on a specimen.

Links to the *Advancing Physics* CD-ROM

Practise with these questions:

10C Comprehension *Visible structures*

50S Short answer *Ants, atoms and chips!*

20E Estimate *Scaling exercises*

Try out these activities:

20S Software-based *Wood and wood products*

50H Home experiment *Sweets and biscuits*

60H Home experiment *Model of a crystal*

Look up these key terms in the A–Z:
Ceramics; composite material; crystals; electron microscopes and atomic microscopy; glass; metals; solid materials

Go further for interest by looking at:

30S Computer screen *Textile images*

Revise using the revision checklist and:

400 OHT *Looking inside glasses*

500 OHT *Looking inside wood*

600 OHT *Looking inside metals and ceramics*

700 OHT *Looking inside polymers*

5.2 Stiff stuff, tough stuff

Glass is notable for two properties: it is transparent and it is brittle – it shatters if you hit it hard enough. The structure of glass explains why it is brittle. Glass – and there are many types of glass – consists of a certain amount of silica together with various metal oxides. For example, soda glass – the stuff that laboratory glassware is made of – contains about 70% silica with a mixture of sodium oxide and calcium oxide. In fact, the two metal oxides give themselves away when you heat a piece of soda glass; note the orangey-yellow colour of the flame. In the flame test for metal ions, sodium is yellow and calcium is brick red.

Glasses are amorphous, or non-crystalline, materials. Their structure in the solid state does not differ much from that in the liquid state – there is little order in the way the atoms are arranged (see top right image on p101). The atoms do not arrange themselves in the regular arrays that are typical of crystalline materials such as metals.

It is this underlying difference in internal structure that explains why glass is brittle and copper is not. In the 1920s, Alan Griffith was working at the Royal Aircraft Establishment at Farnborough when he had an important idea. Calculations showed that most materials ought to be 10 to 1000 times as strong as is actually observed. Griffith argued that the presence of minute cracks and flaws in a material accounted for this discrepancy. They act as stress raisers; in other words, the stress around a crack can be hundreds or even thousands of times as much as the actual applied stress. Bending the glass slightly makes the crack open up and spread through the material. This is how glass is often cut: the glass merchant scratches a fine line, gives the glass one tap and the sheet falls apart exactly along the line.

Griffith's theory is supported by the superior strength of freshly drawn glass fibres, which have few flaws. In a metal, dislocations in the ordered crystalline structure can help to even out high stress concentrations around a crack.

We can think about how glass breaks in terms of the atoms of which it is made and the work which must be done in breaking them apart.

- When you bend a piece of glass it becomes strained elastically. It stores **strain energy**. (This

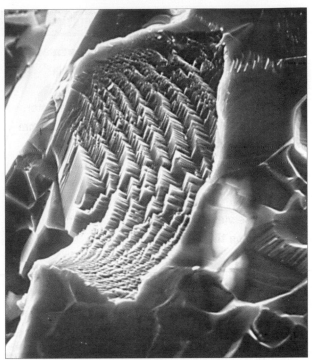

This electron micrograph shows how a crack has moved through a ceramic material – alpha alumina, a form of aluminium oxide. The crack moved diagonally downwards from right to left. It found it easier to travel along horizontal and vertical planes of atoms, and so it has formed a series of steps.

is like the energy stored in a stretched spring.)
- At the tip of the crack, two neighbouring atoms are pulled apart. Work is done in breaking the bond between the atoms.
- Then the next two atoms are pulled apart, and the next two, and so on. The crack moves through the material like a zip being undone. The energy required to do this is known as **fracture energy**.
- Once the piece of glass is broken, it is no longer strained, so it no longer stores strain energy. The energy has been used in breaking the bonds, in sending fragments of glass flying and in making atoms vibrate.

Strength and toughness

Griffith's work on cracks helped to explain the difference between a brittle material like glass and a tough one like steel. It doesn't take much energy to break glass; it takes a lot to break steel. This is illustrated by the figures in the table "Fracture energy and tensile strengths of some common materials" p104.

The toughness of a material is measured by

Fracture energy and tensile strengths of some common materials[*]

Source: *Adapted from J E Gordon 1991 *Structures* (Penguin).

Material	Approximate fracture energy /J m^{-2}	Approximate tensile strength /MPa
glass, pottery	1–10	170
cement, brick, stone	3–40	4
polyester and epoxy resins	100	50
nylon, polythene	1000	150–600
bones, teeth	1000	200
wood	10 000	100
mild steel	100 000–1 000 000	400
high-tensile steel	10 000	1000

Key summary: cracks and stress

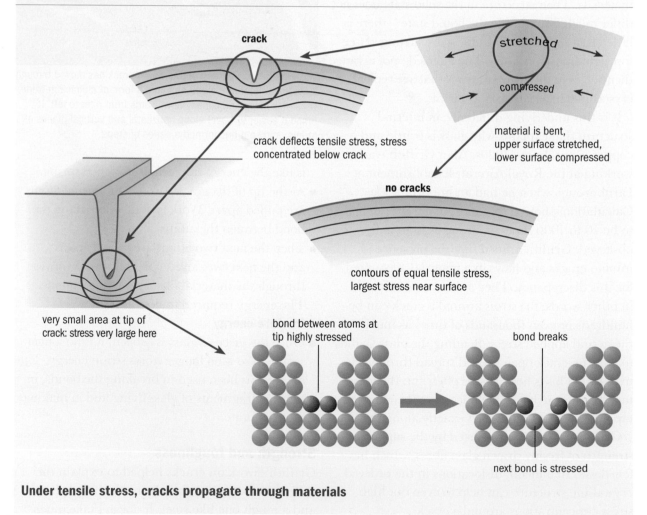

crack

stretched

compressed

crack deflects tensile stress, stress concentrated below crack

material is bent, upper surface stretched, lower surface compressed

no cracks

contours of equal tensile stress, largest stress near surface

very small area at tip of crack: stress very large here

bond between atoms at tip highly stressed

bond breaks

next bond is stressed

Under tensile stress, cracks propagate through materials

the energy needed to deepen and extend cracks, creating a new fractured surface in the cracks. If the energy available from the stresses in the material is larger than the energy needed to extend a crack, the crack will propagate and the material will fail. This is why glass can break almost explosively.

The energy used in fracturing a specimen varies a good deal – the new surface may vary in

Key summary: stopping cracks propagating in metals

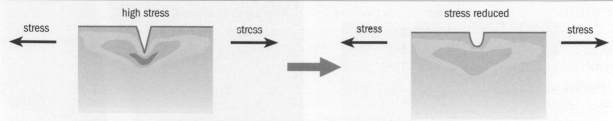

Metals resist cracking because they are ductile. Under stress, cracks are broadened and blunted, they do not propagate.

Metals are tough because they are ductile

Key summary: fracture energy and tensile strength

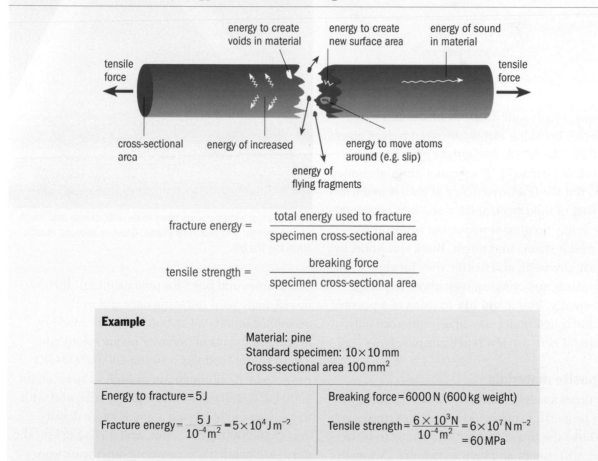

$$\text{fracture energy} = \frac{\text{total energy used to fracture}}{\text{specimen cross-sectional area}}$$

$$\text{tensile strength} = \frac{\text{breaking force}}{\text{specimen cross-sectional area}}$$

Example

Material: pine
Standard specimen: $10 \times 10\,\text{mm}$
Cross-sectional area $100\,\text{mm}^2$

Energy to fracture $= 5\,\text{J}$	Breaking force $= 6000\,\text{N}$ (600 kg weight)
Fracture energy $= \dfrac{5\,\text{J}}{10^{-4}\,\text{m}^2} = 5 \times 10^4\,\text{J\,m}^{-2}$	Tensile strength $= \dfrac{6 \times 10^3\,\text{N}}{10^{-4}\,\text{m}^2} = 6 \times 10^7\,\text{N\,m}^{-2}$ $= 60\,\text{MPa}$

Large fracture energy = tough. Large tensile strength = strong

roughness, the number of small flying pieces can alter, and so on. That is why the figures in the table are labelled "approximate", and are given in very round figures. They do give a rough indication of toughness. You will see from the table that toughness is not at all the same as tensile strength. Glass has quite a high tensile strength but a low toughness: you have to pull hard to break it but it is brittle and snaps without using much energy. A polyester resin is around 10 times as tough as glass, but it breaks at just one-third of the stress figure for glass. You don't have to pull as hard to break it, but it doesn't come apart as readily. It is not as brittle. Materials, then, can be tough and strong, tough

Strong and tough on a log scale

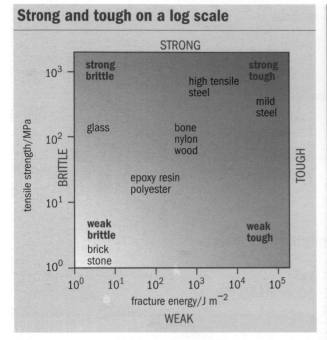

The blades of wind turbines need to be stiff, strong and tough, as well as light and relatively cheap. Glass-reinforced plastic often fills the bill.

but weak, brittle and strong or brittle and weak.

Imagine breaking identically sized rods of glass and steel. The tensile strength of glass is roughly half that of mild steel. They can withstand similar forces. But the fracture energy of glass is much less than that of mild steel, at least one-ten-thousandth of the value. So glass is strong but brittle and mild steel is strong and tough. Brick and stone, by contrast, are weak and brittle: they break at a low tensile stress and come apart without requiring much energy. You might like to think of a possible weak but tough material – fibre-reinforced jelly, perhaps? There are few real examples, however.

Composite materials

Sometimes a single material can not combine all of the properties you would like. For example, you would like the frame of a tennis racquet to be as strong, stiff, tough and light as possible. A way to achieve this is to combine different materials in a composite structure.

One example is glass-reinforced plastic (GRP), which is often used in the hulls of sailing boats, and more recently in the blades of wind turbines.

A similar, higher performing but more expensive composite material is carbon-fibre-reinforced plastic. It is now quite widely used in professional sports equipment (fishing rods, skateboards, golf clubs, tennis racquets, archery bows, mountain-

bike frames and poles for pole vaulting). It is increasingly used in racing cars and in the aerospace industry. Carbon fibres are made by heating filaments of polymer polyacrylonitrile. Their Young modulus is in the 200 to 500 GPa range, which, although not as high as steel (about 1000 GPa), still makes them pretty stiff, and with the advantage of having a much lower density (less than $2000\,\mathrm{kg\,m^{-3}}$). But, and it is a big but, the fibres are brittle. You certainly don't want your pole to snap in the middle of a pole vault.

The way to make a stiff but tough material from carbon fibre is to put the fibres into an epoxy resin matrix. The epoxy resin sticks strongly to the carbon fibres so that they can take up stress. The combination of the two materials has a large Young modulus. A single fibre is still brittle and snaps if a crack propagates across it, but now the fracture damage is limited to just one fibre. The epoxy resin is tough, and when under stress it

A woven mat of carbon fibres before being impregnated with epoxy resin and moulded to the required shape.

The frame of this tennis racquet is made of carbon-fibre reinforced plastic. It is stiff, light and fracture-resistant.

Key summary: fibre composites stop cracks propagating

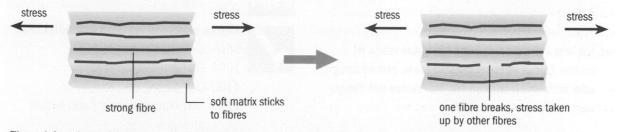

Fibre-reinforced materials use a matrix to share stress among many strong fibres. The matrix also protects the fibres from cracking.

Fibre-reinforced materials are tough because cracks can't propagate through the soft matrix

shares the extra stress among many nearby fibres, so that they are not as likely to break. In this way the composite has the stiffness of the fibres combined with the toughness of the epoxy resin, and the low density of both.

The materials are modern, but the idea is old. "Making bricks without straw" (meaning it's a bad idea) refers to the ancient brick-making practice of adding straw to mud bricks to toughen them. Steel-reinforced concrete works in the same way, so concrete can be used to bear loads in tension as

well as in compression, for example in a cantilever. And of course, biological evolution got there first, "inventing" light, stiff and tough materials such as bamboo and bone.

All of these are excellent examples of where the structure of a material determines its properties. However, now the relevant structure is not at a microscopic or atomic scale, but at a larger, even macroscopic, scale, or at a middling ("mesoscopic") scale in between that of atoms and everyday objects.

Quick check

1. Explain why the stress at the tip of a crack in a brittle material is very large.

2. Explain why bricks can safely be used in buildings, even though brick is a brittle material and has a low tensile strength.

3. Show that the pressure exerted by a 2.5 kg brick when sitting on its largest face (area $20 \times 10^{-3} \, \text{m}^2$) is about 1.25 kPa. Thus show that if the maximum compressive strength of common bricks is 20 MPa, the maximum height of a stack of such bricks, before the bottom gives way, is about 16 km.

4. By quoting data from the table on p104, show which of mild steel and high-tensile steel is the tougher. Which is stronger?

5. Show that a nylon fishing line of diameter 1 mm requires energy of the order of 1 mJ to fracture and that, assuming the average value for the tensile strength of nylon, this corresponds to a load of about 400 N. (For data see the table on p104.)

6. Explain how a composite material made of brittle fibres set in a plastic resin can be tough and stiff, even though the fibres are not tough and the resin is not stiff.

Links to the *Advancing Physics* CD-ROM

Practise with these questions:

40C Comprehension *Photoelastic stress images*

110S Short answer *Bone*

120C Comprehension *Concrete: a material for all seasons*

130C Comprehension *Wire ropes and suspension bridges*

Try out these activities:

130D Demonstration *Photoelastic stress*

300H Home experiment *A jelly composite*

310H Home experiment *Making and testing composite biscuits*

Look up these key terms in the A–Z:

Ceramics; composite material; cracks; glass; materials: properties and uses; mechanical characteristics of materials; metals; solid materials; stretching and breaking

Go further for interest by looking at:

500   OHT *Looking inside wood*

Revise using the revision checklist and:

900   OHT *Cracks and stress*

1000 OHT *Stopping cracks*

1100 OHT *Strong and tough*

1200 OHT *Fracture energy and tensile strength*

5.3 Making more of materials

Gold is a beautiful material. For thousands of years it has been used for making attractive objects – bracelets, medallions, rings, brooches. Today, these may seem to be purely decorative items, but centuries ago to the wearer they were enormously useful, as symbols of power, status and wealth.

Gold was one of the first metals to be worked by man. It is one of the few metals that can be found in its native state, that is, as a pure metal that does not have to be extracted from an ore. Flakes of unoxidised gold can be found in riverbeds, and gold nuggets are extracted from exposed seams in rocks.

There are several ways of working gold into a desired shape. The earliest gold artefacts were made simply by hammering out the metal with hand-held stones. The resulting sheet could then be scratched and pierced to produce the desired effect. A more recently developed technique is rolling, which makes use of the ductility of gold. The metal is passed between a pair of rollers. It is gradually squeezed thinner and thinner; the final gold leaf produced may be only a few hundreds of atoms thick. Gold leaf has been widely used for centuries (and still is today), for example in religious icons, especially by orthodox churches, and for decorating public buildings, such as the domes of mosques.

More elaborate, three-dimensional shapes can be produced by melting gold and pouring it into moulds. (Gold melts at a little over 1000 °C.) A mould is often made by the "lost wax" process. An object is modelled in wax and then wet sand is packed around it. The wax is melted and poured out, leaving a hollow into which molten gold is poured. This technique can produce very finely detailed objects.

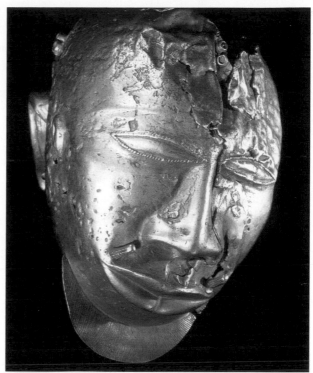

This gold mask was made in the 17th century by the "lost wax" process. It comes from the Ashanti region of present-day Ghana, where west African metal-working techniques had become highly sophisticated.

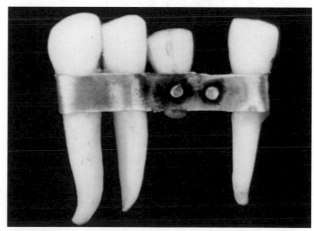

This is the work of an Etruscan dentist 2500 years ago. A band of gold around three of the patient's own teeth is used to hold in place a false tooth, made from a calf's tooth.

A bit of a mouthful

Dentists use about 60 tonnes of gold each year, worldwide. Gold is useful for fillings because it does not corrode or tarnish in the damp environment of the mouth, so you can be proud of your shiny golden smile.

Fillings have to be able to withstand the crushing forces when your jaws close on your food – up to 400 N between the upper and lower molars. So the gold used by dentists is an alloy of gold with other metals such as silver, palladium and platinum. Alloying increases the hardness of a metal.

Gold is a ductile metal. It can be hammered, rolled or drawn out into a desired shape. Ductility is a typical property of metals that makes them different from ceramics and polymers. Think what happens if you try to hammer glass or plastic.

In metals there are mismatches called dislocations in the tidy rows of atoms. It is the

Key summary: shaping and slipping

Atoms in gold are in a regular array: a crystal lattice. To shape the metal, one layer must be made to slide over another.

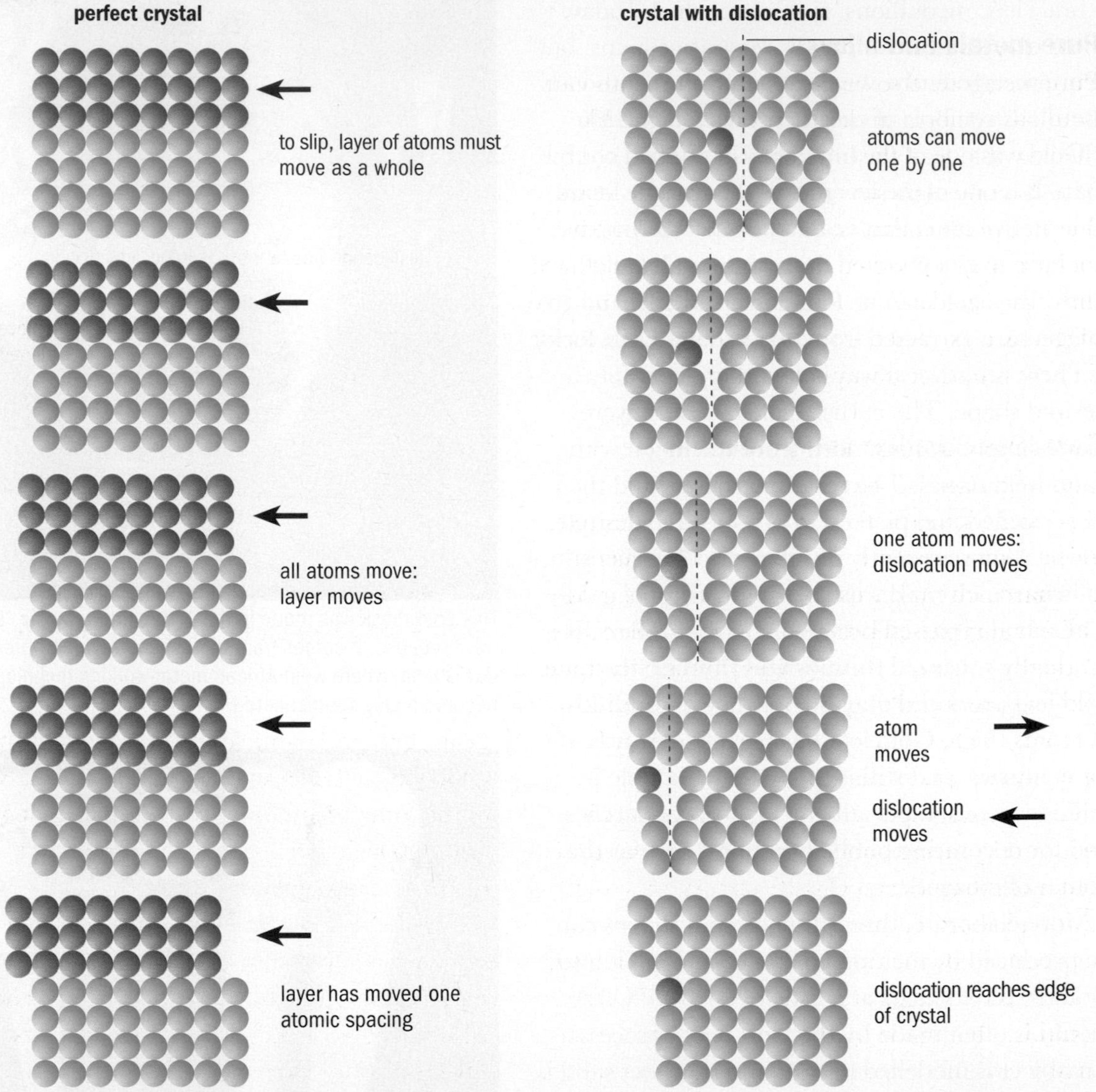

perfect crystal

crystal with dislocation

dislocation

to slip, layer of atoms must move as a whole

atoms can move one by one

one atom moves: dislocation moves

all atoms move: layer moves

atom moves

dislocation moves

layer has moved one atomic spacing

dislocation reaches edge of crystal

in both examples a layer has slipped by one atomic spacing

Wrong model:

Making all the atoms slip together needs considerable energy

This model predicts metals to be 1000 times as strong as they actually are

Better model:

One atom slipping at a time needs much less energy

The dislocation model predicts the strength of metals much better

Dislocations make metals ductile

movement of these dislocations through metal crystals that makes metals ductile. This is a good thing if you are trying to shape a piece of metal, but it is a disadvantage if you want to make a very strong object out of metal. Because of dislocations,

metals may exhibit as little as one-thousandth of the strength that they would otherwise have.

Dislocations can work the other way round too. Flex a copper wire or a paper clip repeatedly between your fingers. This can create dislocations.

When there are many of them, they get entangled and stop each other moving. The metal is now less ductile. It has been **work hardened**.

Pure metals and alloys

Pure metals tend to be very soft. Soft metals can gradually deform under their own weight, like the way that lead flashing used in pitched church roofs becomes thicker at one end over the years due to a phenomenon called creep. Alloying a metal introduces other elements that usually have different sized atoms. This disrupts the regularity of the metal's crystal structure and makes it harder for atoms to slide past one another.

Ceramics, familiar and unfamiliar

Ceramic materials have been around for thousands of years. Pottery is one such example, and it illustrates the typical properties of ceramics. It is hard, stiff and wear resistant. It can survive at fairly high temperatures and resists corrosion. Pottery also shows the main weakness of ceramics, which are generally brittle, so they can't withstand impacts. That's why archaeologists usually find only broken bits of ancient pots.

Other traditional ceramic materials include brick and concrete, and there are many natural materials that we can classify as ceramics, such as flint and stone. Ceramics are crystal structures, generally built from ions of two or more elements.

Increasingly, engineers are turning to newer ceramics. Pottery is made from clay, which can have a variable composition, but engineering ceramics are made from highly purified starting materials, so that their composition and structure can be carefully controlled. From the names of some of these materials – aluminium oxide, silicon nitride, silicon carbide – you can see that they are relatively simple compounds in which the atoms are strongly bonded together.

Fatal flaws

Besides brittleness, there is another big problem with ceramics: they are difficult to shape. It's easy to understand why this is: it's because of their hardness. They can't be melted and shaped like most metals and polymers. So ceramic objects are often made by shaping the raw materials and then

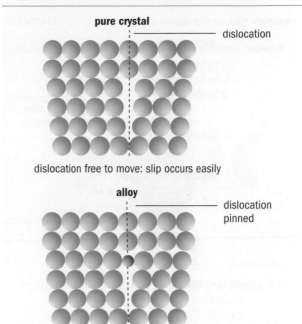

Key summary: metals and metal alloys

pure crystal

dislocation

dislocation free to move: slip occurs easily

alloy

dislocation pinned

alloy atom pins dislocation: slip is more difficult

Alloys are generally less ductile than pure metals

heating them. Think about how a cup is made.

Clay is shaped and then fired in a kiln. Some of the minerals in the clay melt to form a glassy liquid, which, when cooled and solidified, "glues" the material together.

Engineering ceramics may be processed in the same way. Silicon powder is moulded into the shape of, say, a turbine blade. When it is heated with nitrogen, a chemical reaction occurs and silicon nitride is formed. An alternative approach is sintering. Silicon nitride powder is placed in a mould and heated under high pressure. The surfaces of the particles melt and flow, so that they become glued together.

The key to making a good ceramic object is to avoid flaws in its structure. Any tiny cracks, holes or impurities can result in a fatal weakness. Just as with glass, a crack can propagate rapidly through a ceramic material, and this explains why ceramics are generally weaker in tension than in compression.

It may be possible to have the best of both worlds by combining ceramics and metals. A "cermet" is a combination of ceramic particles in a metal matrix (see image middle left on p98). Such a material can be hard, tough and strong.

Key summary: ceramics versus metals

Ceramics have rigid structures

Covalent structures for example silica, diamond and carborundum

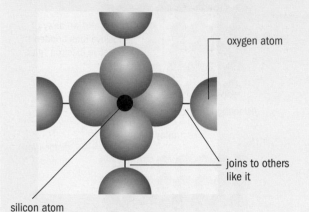

oxygen atom

joins to others like it

silicon atom

The atoms are linked in a rigid giant structure

Atoms share electrons with neighbouring atoms to form covalent bonds. These bonds are directional: they lock atoms in place, like scaffolding.

The bonds are strong: silica is stiff

The atoms cannot slip: silica is hard and brittle

Ionic structures for example common salt

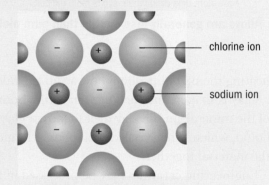

chlorine ion

sodium ion

The ions are linked in a rigid giant structure

Some atoms give electrons to other atoms to form an ionic bond. Because like charges repel and unlike charges attract, the charged ions hold each other in place.

The bonds are strong: salt crystals are stiff

The ions cannot slip: salt crystals are hard and brittle

Metals have non-directional bonds

Metallic structures for example gold

negative electron "glue"

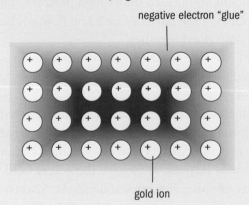

gold ion

Atoms in metals are ionised. The free electrons move between the ions. The negative charge of the electrons "glues" the ions together, but the ions can easily change places.

The bonds are strong: metals are stiff

The ions can slip: metals are ductile and tough

The ions are held together but can move

Key summary: explaining stiffness and elasticity

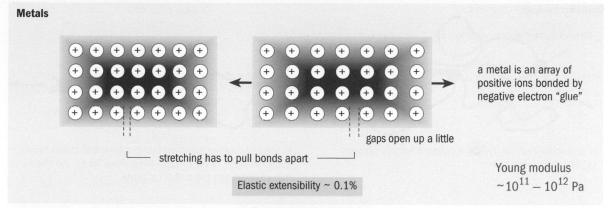

Metals

a metal is an array of positive ions bonded by negative electron "glue"

gaps open up a little

stretching has to pull bonds apart

Elastic extensibility ~ 0.1%

Young modulus
~$10^{11} - 10^{12}$ Pa

Stretching a metal stretches bonds – but not by much

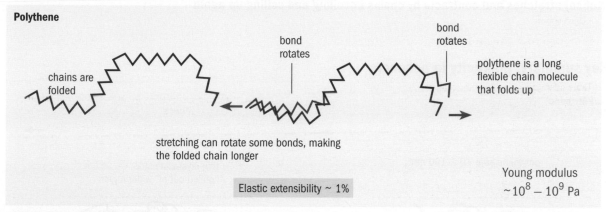

Polythene

chains are folded

bond rotates

bond rotates

polythene is a long flexible chain molecule that folds up

stretching can rotate some bonds, making the folded chain longer

Elastic extensibility ~ 1%

Young modulus
~$10^{8} - 10^{9}$ Pa

Stretching polythene rotates bonds

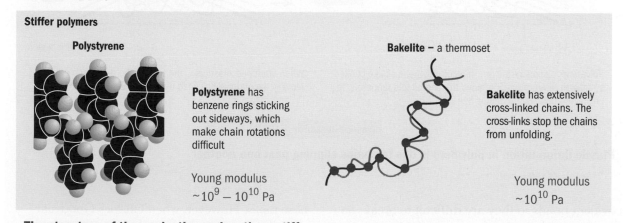

Stiffer polymers

Polystyrene

Bakelite – a thermoset

Polystyrene has benzene rings sticking out sideways, which make chain rotations difficult

Bakelite has extensively cross-linked chains. The cross-links stop the chains from unfolding.

Young modulus
~$10^{9} - 10^{10}$ Pa

Young modulus
~10^{10} Pa

The structure of these plastics makes them stiff

Fibres and fabrics, polymers and plastics

Polymers are substances made from long-chain molecules – polythene, nylon, Perspex. These are synthetic polymers, invented in the 20th century. There are many natural polymers too – wool, cotton, leather. Many biological molecules are polymers – starch, proteins and, of course, DNA.

Polymers tend to be flexible but strong – that's why most of our clothes are made of polymers.

A molecule of polythene – more correctly, poly(ethene) – is a long chain of identical, repeating units. A molecule like this is very floppy because it is free to rotate about its bonds. However, the atoms are joined together by strong,

Key summary: rubber

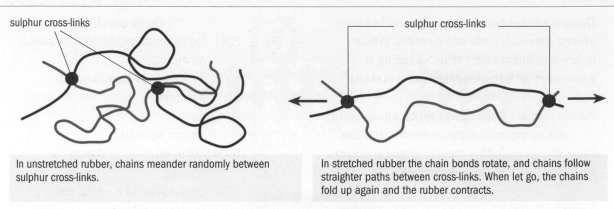

sulphur cross-links

sulphur cross-links

In unstretched rubber, chains meander randomly between sulphur cross-links.

In stretched rubber the chain bonds rotate, and chains follow straighter paths between cross-links. When let go, the chains fold up again and the rubber contracts.

Elastic extensibility > 100%

Rubber stretches and contracts by chains uncoiling and coiling up again

Key summary: plasticity in polymers

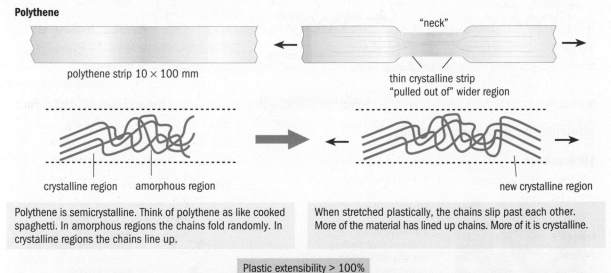

Polythene

polythene strip 10 × 100 mm

"neck"

thin crystalline strip "pulled out of" wider region

crystalline region   amorphous region

new crystalline region

Polythene is semicrystalline. Think of polythene as like cooked spaghetti. In amorphous regions the chains fold randomly. In crystalline regions the chains line up.

When stretched plastically, the chains slip past each other. More of the material has lined up chains. More of it is crystalline.

Plastic extensibility > 100%

Plastic deformation in polymers is due to chains slipping past one another

covalent bonds so they are difficult to break. So polythene is flexible but strong.

Rubber control

Rubber latex is a runny white liquid – the sap of the rubber tree, which is a native of South America. Christopher Columbus learned from the indigenous people there how to make heavy black rubber balls. At first rubber was regarded simply as a curiosity, an amusing, sticky, bouncy substance, useful for waterproofing clothing and

hose pipes. However, when Charles Goodyear invented the process of vulcanisation in 1839, rubber suddenly became a valuable material. To vulcanise rubber, it is heated with sulphur. Cross-links are formed between the polymer chains by the sulphur atoms. The more sulphur you add, the more cross-links form and the stiffer the rubber becomes. Control of the structure of the material gave control over its properties. So rubber found many new applications, in tyres, shock absorbers and electrical insulation to name just a few.

Quick check

1. There are three types of strong bond between atoms: covalent, ionic and metallic. Which is/are non-directional? Which bonding is associated with the toughness of a material?

2. Metals contain dislocations which allow layers of atoms to slip past each other with relative ease. What aesthetic property of metal objects does this fact help to make possible?

3. A piece of rubber 6 cm long cut from a rubber band will easily extend to 7 cm when supporting a small load. Bonds between atoms typically break when the distance between the atoms is increased by 1%. Show that the extension of the rubber cannot simply be due to stretching of the bonds between atoms.

4. If the maximum elastic strain of a polymer is 1%, and its Young modulus is 10^8 Pa, show that the corresponding stress is 10^6 Pa.

5. Molecules of DNA have been stretched using "optical tweezers". Using the following data, show that the Young modulus of DNA is 10^8 Pa. Load $= 400 \times 10^{-12}$ N; resulting strain $= 20\%$; cross-sectional area of a DNA strand $= 2 \times 10^{-17}$ m^2.

6. Hooke's law states that extension Δx is proportional to load F and is usually written $F = k\Delta x$. The Young modulus $E = \dfrac{\text{stress}}{\text{strain}} = \dfrac{F/A}{\Delta x/L}$. Show that the Hooke's law constant k is related to the Young modulus E by the relationship $k = \dfrac{EA}{L}$.

Links to the *Advancing Physics* CD-ROM

Practise with these questions:

70X   Explanation–exposition *Questions on metals*

100S   Short answer *Questions on polymers*

Try out these activities:

160H   Home experiment *Making ice crystals*

260D   Demonstration *A model for stretching rubber*

Look up these key terms in the A–Z:
Ceramics; metals; materials: properties and uses; mechanical characteristics of materials; polymers; solid materials; stretching and breaking

Go further for interest by looking at:
240S   Computer screen *Materials: Yesterday and tomorrow*

Revise using the revision checklist and:
1600   OHT *Shaping and slipping*
1700   OHT *Metals and metal alloys*
1800   OHT *Ceramics versus metals*
1900   OHT *Explaining stiffness and elasticity*
2000   OHT *Plasticity in polythene*
2100   OHT *Elastic behaviour in rubber*

5.4 Controlling conductivity

Semiconductors are fascinating materials, and the technology that uses them has transformed our world. They are electrical conductors, but their conductivity is much less than that of metals. Silicon and germanium, two examples of semiconductors, lie in the middle of the periodic table between the metals on the left and the insulators on the right. Because of semiconductors' importance in modern electronic devices, an industry has developed to produce extremely pure semiconducting materials, so that silicon crystals are available with 99.9999999% purity – just one impurity atom among a billion silicon atoms.

Transistors are the semiconductor devices that make possible our personal computers, hi-fi systems, body scanners, aircraft-control systems, telephone networks etc. Transistors are used to control and amplify electrical signals; the latest microprocessor at the heart of a personal computer contains tens of millions of them, all packaged in a single integrated circuit or chip. The task of producing the detailed design of such a chip is beyond any human, so computers design each new generation of chips according to instructions laid down by a team of engineers – so we have chips designing chips.

Birth of the information age

The first transistor was demonstrated in 1947. It was invented by three physicist–engineers working at the research labs of the Bell Telephone company: John Bardeen, Walter Brattain and William Shockley. Their device amplified a small electric current by a factor of almost 100. Until this time, electrical amplification was done using valves, which are glass vacuum tubes through which electric currents flow. Valves still have uses today, but they are bulky and take time to warm up. The attraction of the transistor is that it is a solid-state device – it is entirely made from solid materials – so it is much more robust and reliable than a valve.

A metal conducts less well when heated because the vibrating ions scatter the moving electrons. This scattering effect occurs in semiconductors too, but it is masked by a bigger effect, the great increase in the number of free electrons, which have broken away from their atoms.

The first transistor was a ramshackle affair, made from a small piece of germanium. The "business end" is the point where the triangle of plastic presses a split piece of gold foil down on to the horizontal slab of germanium. Today, many millions of transistors are integrated into a single semiconductor chip.

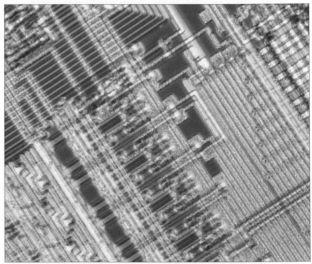

This integrated circuit has been made by imprinting microscopic electronic components onto the surface of a wafer of silicon. Magnification: × 100 at 35mm size.

Objects made of materials that you generally think of as insulators will often conduct a tiny current. This may be because of moisture on the surface: not really a property of the material, but important, especially with glass insulators. Making an insulating material hot will often make

Key summary: conduction by metals and semiconductors

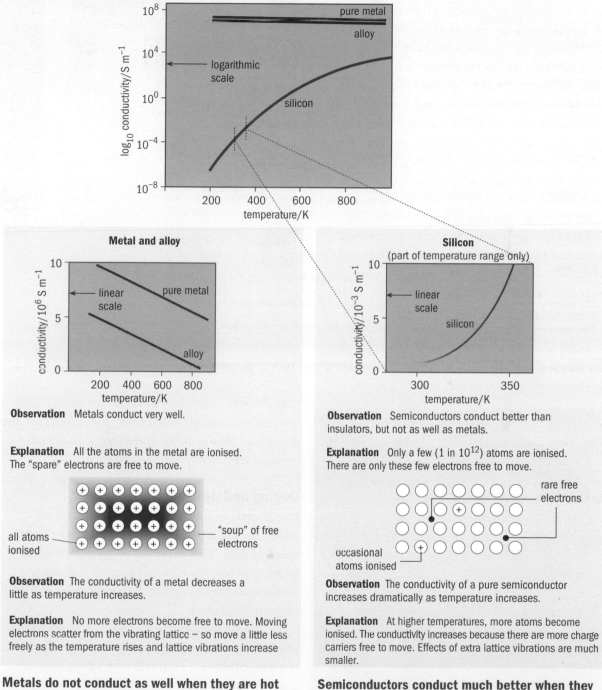

Metal and alloy

Observation Metals conduct very well.

Explanation All the atoms in the metal are ionised. The "spare" electrons are free to move.

all atoms ionised

"soup" of free electrons

Observation The conductivity of a metal decreases a little as temperature increases.

Explanation No more electrons become free to move. Moving electrons scatter from the vibrating lattice – so move a little less freely as the temperature rises and lattice vibrations increase

Metals do not conduct as well when they are hot because charge carriers become less mobile, but their number stays the same

Silicon (part of temperature range only)

Observation Semiconductors conduct better than insulators, but not as well as metals.

Explanation Only a few (1 in 10^{12}) atoms are ionised. There are only these few electrons free to move.

rare free electrons

occasional atoms ionised

Observation The conductivity of a pure semiconductor increases dramatically as temperature increases.

Explanation At higher temperatures, more atoms become ionised. The conductivity increases because there are more charge carriers free to move. Effects of extra lattice vibrations are much smaller.

Semiconductors conduct much better when they are hot. More charge carriers are freed and they only become a little less mobile.

it conduct a little better, by freeing charge carriers to move. For example, glass conducts quite well when it is red-hot. Sodium and other ions in the glass can get enough energy to hop about in the amorphous structure and carry a current.

The structure of a material can be the reason for a number of its properties. For example, the free electrons in a metal (a) glue the ions together with non-directional bonds making the metal strong but ductile, (b) allow the metal to conduct electricity

Key summary: properties that go together

Property	Reason
Metals	*because* *the free electrons in metals:*
conduct well	are mobile and so carry electric current
are shiny	oscillate in light, scattering light photons
are stiff	"glue" ions together strongly
are ductile	provide a non-directional "glue", letting ions slip
Ceramics	*because* *the ionic or covalent bonds holding them together:*
are insulators	lock electrons to ions or atoms, with none free to move
are stiff	are strong bonds, hard to stretch
are brittle	are directional bonds, so that atoms or ions cannot slip
Polymers	*because* *the covalent bonds stringing monomers in long chains:*
are insulators	lock electrons to atoms, with none free to move
are often flexible	can rotate, letting chains stretch or fold
are often plastic	make chains that can slip past one another

The bonding and structure of a material explain whole sets of properties

This ancient Roman mirror is made of silver and has a polished convex face for viewing.

ancient civilisations could make metal mirrors and shiny gold masks (p109) is that these materials can conduct electricity.

Doping and devices

Integrated circuits are made in "clean rooms", where the air is purified to remove almost every speck of dust. This is essential, because to make a chip you have to control the purity of the semiconducting material. Any impurities will undo all your hard work.

A chip is a miniaturised electric circuit, with insulating, conducting and semiconducting regions. Making a chip relies on being able to control the conductivity of silicon. To make an insulating region, silicon can be exposed to oxygen or nitrogen to produce silicon oxide or silicon nitride. To produce a conducting region, it may be coated with gold. But the vital step that produces the semiconducting parts of the transistors is the process called doping. Ions are sprayed onto an exposed surface of a silicon chip and are absorbed into its crystalline structure. Depending on the type of ions used, this produces

well, and (c) make the metal shiny by absorbing and re-emitting light that strikes its surface. Three seemingly distinct properties have a common origin. So it's a striking fact that the reason why

Key summary: conduction in doped silicon

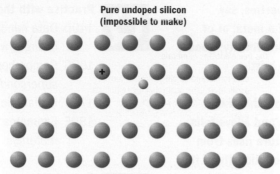

**Pure undoped silicon
(impossible to make)**

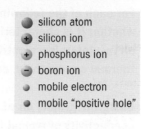

- silicon atom
- silicon ion
- phosphorus ion
- boron ion
- mobile electron
- mobile "positive hole"

Less than one in a million million silicon atoms are ionised,
giving a very small fraction of electrons that are free to move

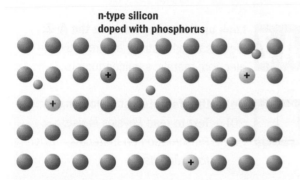

**n-type silicon
doped with phosphorus**

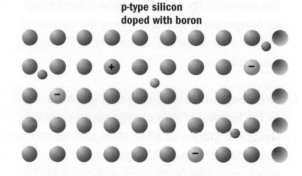

**p-type silicon
doped with boron**

The phosphorus atoms ionise, giving electrons free to move
throughout the material

The boron atoms ionise, taking electrons from silicon atoms
and leaving "positive holes" free to move throughout the material

$$
\begin{array}{c}
\text{Si} \\
\bullet\bullet \\
\text{Si} \bullet\bullet \text{P}^+ \bullet\bullet \text{Si} \\
\bullet\bullet \\
\text{Si}
\end{array}
\quad \nearrow \bullet \quad
\begin{array}{l}
\text{spare electron} \\
\text{free to conduct}
\end{array}
$$

Phosphorus has five electrons in its outer shell. Four are shared
with silicon atoms. One becomes free to move and conduct, leaving
positive phosphorus ions.

n-type: electrons conduct

$$
\begin{array}{c}
\text{Si} \\
\bullet\bullet \\
\text{Si} \bullet\bullet \text{B}^- \bullet\bullet \text{Si} \\
\bullet\bullet \\
\text{Si}
\end{array}
\quad \swarrow \bullet \quad
\begin{array}{l}
\text{stolen electron} \\
\text{leaves mobile hole}
\end{array}
$$

Boron has three electrons in its outer shell. One more is "stolen"
to give four to share with silicon atoms. The "electron hole" left
behaves like a mobile positive charge.

p-type: holes conduct

either n-type or p-type silicon.
- In n-type silicon the impurities provide
  extra electrons that increase the material's
  conductivity.
- In p-type silicon the impurities provide "holes"
  that behave like positively charged conduction
  electrons. A current flows when holes move from
  positive to negative.

The whole chip is built up in layers. Doping ions

of both kinds are implanted through masks that
place them where they are needed. Then areas
may be removed by etching once the regions
that are to be retained are protected, again using
a mask. Conducting gold or aluminium tracks
are laid down to provide electrical connections.
Layer by layer, under computer control, the whole
complicated chip circuitry is constructed.

See chapter 2 for uses of microcircuits in sensors.

Quick check

1. For each of the following properties, say whether it is characteristic of a metal or of an insulator: low electrical conductivity; good thermal conductivity; ductile; brittle.

2. Using the graphs on p117, show that the conductivity of typical metals and alloys falls by about 10% as the temperature rises from 300 K to 350 K whereas the conductivity of the semiconductor silicon rises by about 900%.

3. Although both graphs on p117 that display the conductivity of silicon show it increasing as temperature increases, they are curving in opposite directions. Explain why this is not a contradiction.

4. The electrical conductivity σ is proportional to the number of free charge carriers per m^3. Using the graphs on p117 show that metals have about twice as many charge carriers per m^3 when compared to an alloy, and about 10^{11} as many as semiconducting silicon.

5. In the periodic table of elements, phosphorus lies to the right of silicon. Will doping silicon with phosphorus produce a p-type or n-type material?

6. Suppose you are given data on the conductivity of a metal and of a semiconductor over a wide range of temperatures, and are asked to plot conductivity against temperature. Sketch the shapes of the graphs you would expect. Would you use a linear or logarithmic scale for conductivity?

Links to the *Advancing Physics* CD-ROM

Practise with these questions:

140D Data handling *How resistivity changes with temperature*

160C Comprehension *High-temperature superconductivity*

170X Explanation–exposition *Conductivity*

180E Estimates *Estimating with materials*

Try out this activity:

320E Experiment *Calibration of a thermistor*

Look up these key terms in the A–Z:

Electrical conductivity and resistivity; electron; metals; semiconductors

Go further for interest by looking at:

20T Text to read *Physics in use: Presentation on materials – Briefing for students*

Revise using the revision checklist and:

2200 OHT *Conduction by metals and semiconductors*

2300 OHT *Free electron model of metal*

2500 OHT *Effect of temperature on conductivity*

2600 OHT *Conduction in doped silicon*

Summary check-up

Structures ✓

- Understanding the structure of a material can help us to understand its properties; by altering the structure we can control the properties
- Metals generally have an ordered, polycrystalline structure
- Glasses have an amorphous structure, similar to that of a liquid

Bonding ✓

- Insulators (such as ceramic materials) are generally held together by strong, directional bonds (ionic and covalent)
- Metals are held together by strong, non-directional bonds (metallic bonding)
- Polymer molecules have a long-chain structure, with strong bonds within the chains

Mechanical properties ✓

- When a metal or ceramic is stretched elastically, the bonds between neighbouring atoms are extended very slightly
- When a polymer is stretched elastically, the atoms rotate about their bonds
- Metals are ductile because of the presence of dislocations that can move through the material
- Ceramics are brittle because cracks and other flaws limit their strength
- Composite materials can be strong and tough by limiting crack propagation in the brittle component

Electrical properties ✓

- Metals are good conductors; ceramics and polymers are generally good insulators
- Semiconductors have intermediate conductivity
- The conductivity of metals decreases gradually with increasing temperature, while the conductivity of semiconductors increases rapidly
- The conductivity of a semiconductor can be increased by doping with atoms of other elements nearby in the periodic table

Questions

1. **Materials can be classified in a variety of ways. Here are some examples of classes: metals, polymers, ceramics and semiconductors. For each of the statements that follow, say which one or more of these classes it might relate to.**
   (a) High electrical conductivity because of the presence of many free electrons.
   (b) Electrical conductivity increases rapidly with increasing temperature as the number of free electrons increases.
   (c) Non-directional bonds result in ductility.
   (d) May be capable of large elastic deformation as atoms rotate about bonds.
   (e) Can be weak due to the presence of surface cracks.

2. **We can predict how strong a material might be if we know how strong the bonds between neighbouring atoms are. However, most materials turn out to be much weaker than our predictions.**
   (a) Explain why glass is strong but brittle.
   (b) Explain why metals are generally weaker than predicted.
   (c) Glass-fibre-reinforced plastics are strong materials. Explain how fibre reinforcement increases the strength of plastic.

3. **Use the data on p104 to estimate:**
   (a) the tensile force needed to stretch and break a wooden metre rule;
   (b) the tensile force needed to break a tooth;
   (c) the tensile force needed to break a nylon fishing line as thin as a hair;
   (d) the tensile force needed to break a 5 mm diameter high-tensile steel cable.

4. **It is possible to break a paper clip by repeatedly flexing the wire backwards and forwards.**
   (a) Describe what you would feel as you did this.
   (b) Explain in terms of the changing internal structure of the wire why this happens.

5. **The graph shows how the resistivity of each of three materials changes as the temperature increases. The three materials are: a pure metal; a pure semiconductor; a doped semiconductor.**

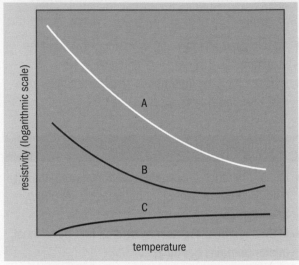

   (a) Say which line on the graph corresponds to which material.
   (b) Explain your choices.

6. **Plastic bags, such as those used for storing food or holding rubbish, can be permanently stretched quite a lot before the material tears. Rubber bands stretch easily but relax back to their original length.**
   (a) Explain the behaviour of plastics such as polythene, in terms of the behaviour of molecules.
   (b) Explain the behaviour of rubber in terms of the behaviour of molecules.

7. **Look at the table of strength and fracture energy on p104, and consult the CD-ROM Materials Database.**
   (a) How much tougher is mild steel than high-tensile steel (an alloy)?
   (b) Explain the difference in terms of their internal structures.
   (c) For what kinds of uses is mild steel to be preferred to high-tensile steel?

6 Wave behaviour

Think of the rainbow reflections from a compact disc or of the colours in a butterfly's wings. Think of a melody played on a solo instrument or the beauty of the song of a blackbird. All have to do with how waves behave. In this chapter we tell the story of how ideas about waves and the nature of light have changed over the centuries. We will look at:

- colours and sounds produced when waves "add together"
- answers to the question, "what is light?"
- effects of interference and diffraction
- and how to represent waves using spinning arrows called phasors

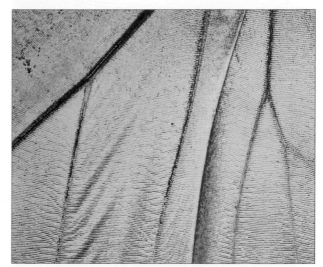

The wing of Brazilian butterfly *Morpho sulkowski* is covered in microscopic thin transparent platelets. These "tiles" are just the right thickness and are laid out on the surface in just the right way to reflect back blue light at certain angles.

6.1 Beautiful colours, wonderful sounds

The beautiful shimmering blue wing of the *Morpho* butterfly is an illusion: it is not really blue at all. Seen from most angles it is just a dull brown or grey colour. But at a particular angle the wing only reflects back blue light. All of the other colours in white light are removed.

The delicate colours of semiprecious stone opal are due to light scattering from regular arrays of tiny spheres of silica. Look at light reflected by the silvery surface of a compact disc: again you see colours – red, yellow, green, blue. This time the colours come from light scattered at the equally spaced rows of the finely spaced spiral of "dots" on the disc (chapter 3, p60). Light reaching your eye from an opal or from a compact disc comes from many rows or lines on the surface, all equally spaced. Light scattered to your eye from a particular row travels further than that from the row in front and thus takes longer. If this time lag brings wave crests and troughs together, light of that colour is weakened; if it brings a crest together with a later crest that colour is strengthened.

The next time it rains, go out and look for oil films on puddles. You can often see a shifting

A close-up of the surface of an opal. Light is diffracted from regular arrays of spheres of silica in the material's structure.

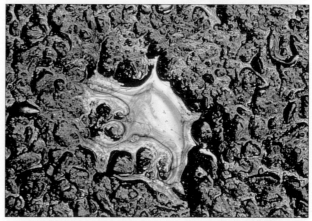

An oil slick on a puddle. The colours come from combining the light reflected from the front and the back of the oil film.

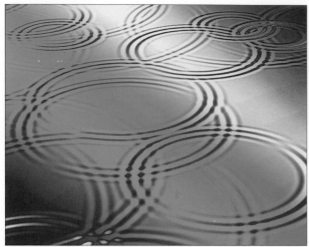
Water ripples move through one another.

In this stormy sea you can see little waves on top of big ones.

A famous Japanese print, *The Wave*, shows a stylised breaker.

pattern of colours on the surface of these films – especially if the Sun is out. This time the colours are produced by light coming to your eye from both the front and the back of the oil film, with the light from the back having to travel a bit farther than the light from the front. Colours are again strengthened or weakened according to whether

the time delays bring crests together with crests or with troughs. Soap bubbles show similar colours to oil slicks; again light reflects from both the inner and outer surfaces of the bubble.

All of these colours have something remarkable in common – they shimmer and change as you move your eye. This is because, as your eye moves, the paths taken by parts of the light change in length. This changes the time lags between the waves so that different colours are strengthened or weakened. Think, for example, of the iridescent plumage on the neck of a Mallard duck. By contrast, the colour of a shirt or blouse doesn't change if you see it from different angles. Ask yourself, where are these different kinds of colour made – in your eye, in the coloured object, or in neither? It's a subtle question.

These colours come from light combining with light. So how can white light plus white light make blue light, or red, or green? The answer lies in the different paths the light takes. These path differences decide by how much the light waves of a particular colour are in or out of step with one another, and so how they combine when they come together, perhaps in your eye.

Superposition: waves on top of one another

One day in 1933, the crew members of the US ship *Ramapo* crossing the North Pacific were horrified to see a wave of more than 30 metres high bearing down on them. There are many such stories about the sudden appearance of gigantic ocean waves, apparently from nowhere. These rare and terrifying monsters – known to marine insurers as "100-year waves" – arise from the chance meeting of many waves that just happen to pile on top of one another at a given moment. The opposite can happen too, a moment of calm as waves that are peaking meet waves that are sinking. Crews of ocean racing yachts often wait for such a "flat" before altering the boat's course.

Waves on water simply pass right through one another. They aren't like moving objects that can hit one another and bounce. They are just movements of the water's surface, and the motion from one wave just adds onto the motion from another. Like snakes in a snake pit, waves crisscross one another without noticing that others are there.

Key summary: phase and angle

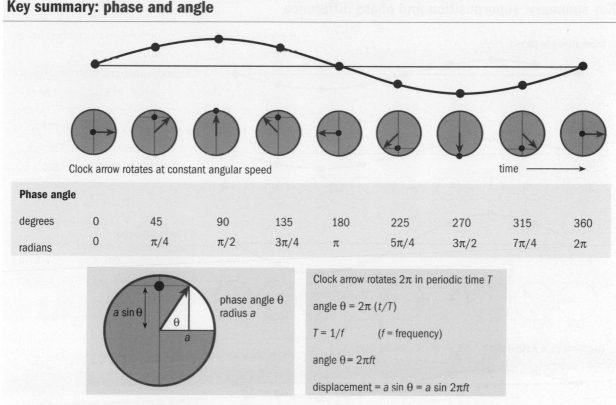

Clock arrow rotates at constant angular speed

time ⟶

Phase angle

degrees	0	45	90	135	180	225	270	315	360
radians	0	$\pi/4$	$\pi/2$	$3\pi/4$	π	$5\pi/4$	$3\pi/2$	$7\pi/4$	2π

$a \sin\theta$

phase angle θ
radius a

θ

a

Clock arrow rotates 2π in periodic time T

angle $\theta = 2\pi\,(t/T)$

$T = 1/f$ (f = frequency)

angle $\theta = 2\pi f t$

displacement $= a \sin\theta = a \sin 2\pi f t$

The phase at a point on a wave can be indicated by the angle of a rotating arrow (phasor)

The same idea lies behind thinking of a complex sound as a spectrum of different frequencies all sounding together, as in chapter 3.

The idea that motions from different waves all pile on top of one another at a given place is called the **principle of superposition**. Superposition just means "placing on top". But because waves go down as well as up, the effect of combining the motion of two waves is not always a bigger motion; it can be a smaller one, too. In adding waves, more can mean less, as well as more.

It's worth pointing out that the simple adding together of waves doesn't always work. On a rough day at sea, high wave crests break making "white horses". The superposition is not linear: the motions don't simply add to one another. But for light and radio waves the principle works extremely well.

Phase differences

When two waves come together at a place, the result depends on what the two waves are doing. If they are going up and down together, in step with one another, the result is an oscillation with an

amplitude equal to the sum of the two. But what if one wave is going up as the other goes down – if they are out of step? Then the resulting oscillation is the difference of the two, so can be smaller than either original wave. It even makes a momentary "flat" if the two waves are equal in amplitude.

The word **phase** is often used to describe a stage in a change that cycles round, for example the phases of the Moon. The phase of a wave motion says where it is in its wave cycle. Two waves doing the same things at the same moment are said to be **in phase**. They have no **phase difference**. Two waves doing exactly opposite things at the same moment are said to be **out of phase** or **in antiphase**. These aren't the only possibilities. One wave may be only just a bit behind or ahead of the other, with a small phase difference between them.

Phases or phase differences are simple to measure. They are just angles. A rotating arrow called a **phasor** is used to keep track of where a wave is in its cycle. This arrow "clock" turns right round once as the wave goes through one whole cycle; it then repeats a new cycle. Between two

Key summary: superposition and phase difference

Oscillations in phase

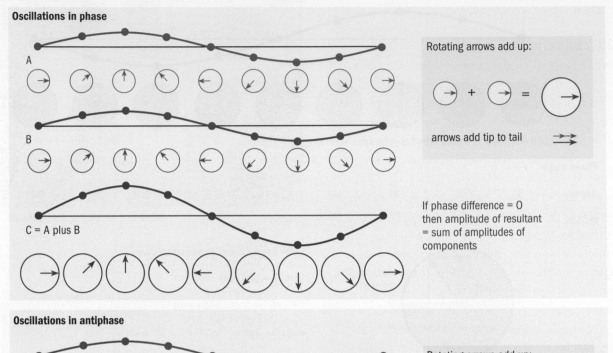

Rotating arrows add up:

If phase difference = 0
then amplitude of resultant
= sum of amplitudes of
components

Oscillations in antiphase

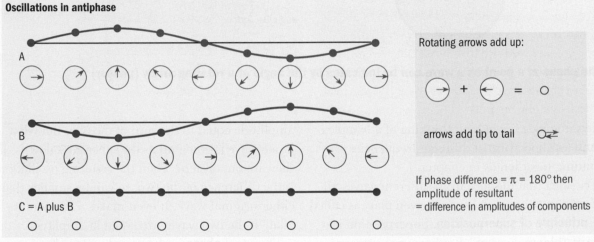

Rotating arrows add up:

If phase difference = π = 180° then
amplitude of resultant
= difference in amplitudes of components

Oscillations with 90° phase difference

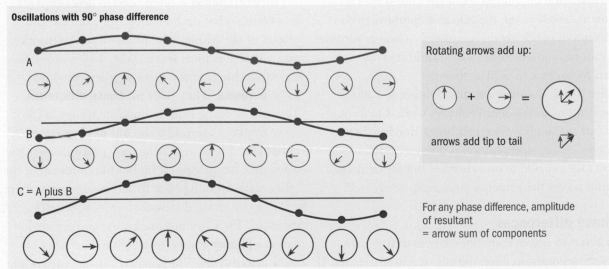

Rotating arrows add up:

For any phase difference, amplitude
of resultant
= arrow sum of components

Wave motions can be combined by adding phasors at each point

waves that are in phase, the difference in phasor angle is zero. Between two waves in antiphase, the angle it is half a complete circle, 180° or π radians.

Oil- and soap-film colours

Superposition explains the colours that you can see in soap bubbles and on oily water. Light falls on the film: some of it is reflected back from the top surface, some passes into the film and reflects from the lower surface. The light that goes into the film travels farther and goes more slowly than the light that comes straight back. There is a delay between light from the front and back surfaces. Because of the delay, peaks and troughs of the two light beams can fall on one another. Superposition says that, if this happens, then no light comes back and the film looks dark.

Usually, you view soap films or oil slicks in daylight, which is a spectrum of different frequencies. If the thickness of the film is just right to remove, say, red light, it won't remove other colours and the film looks blue-green. At another angle, the difference in paths may be just right to remove green light; there the film looks reddish-blue or purplish. This is why soap and oil films show secondary colours, and why the colours shift as the film changes thickness and as you move your head.

The same idea, that light that has travelled different distances can be out of step and cancel, is used in the way binary digits are read from a CD-ROM (see chapter 3, p60).

In all these cases it is the time delay between two parts of the same wave front which matters. For a peak to be delayed so that it falls on a trough of the other part of the wave, it must travel an extra distance equal to half a wavelength of the light in the material of the thin film. Or, of course, any odd number of half-wavelengths. In addition, there may be phase changes at the reflecting surfaces, with peaks coming back as troughs, and vice versa.

An important use for thin films is in making better lenses. Modern camera and binocular lenses are coated with thin surface layers that prevent reflection of light in the middle of the visible spectrum. Such "bloomed" lenses often have a purplish appearance.

Superposition explains the colours that you see in soap bubbles.

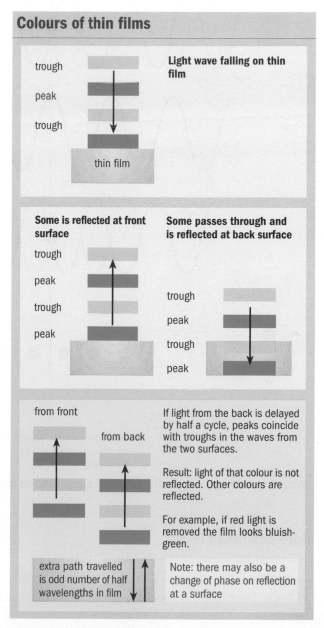

Colours of thin films

Light wave falling on thin film

trough
peak
trough
thin film

Some is reflected at front surface

trough
peak
trough
peak

Some passes through and is reflected at back surface

trough
peak
trough
peak

from front   from back

If light from the back is delayed by half a cycle, peaks coincide with troughs in the waves from the two surfaces.

Result: light of that colour is not reflected. Other colours are reflected.

For example, if red light is removed the film looks bluish-green.

extra path travelled is odd number of half wavelengths in film

Note: there may also be a change of phase on reflection at a surface

Key summary: standing waves

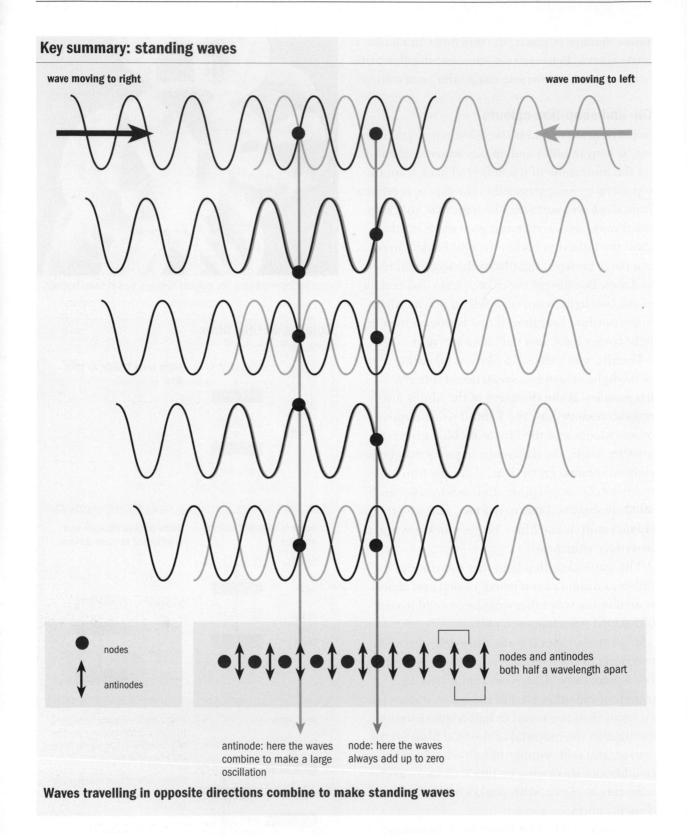

Waves travelling in opposite directions combine to make standing waves

More can mean less

Light plus light can give less light. This happens when two beams of light come together but are out of phase. Superposition here leads to destructive interference. The name interference is a bit misleading, because the whole thing works by the waves not taking the slightest notice of one another, but just adding together to give very little. However, the name is very common and you will meet it frequently. Of course, when waves are in phase, light plus light gives even more light. This is also called interference (constructive interference).

It is not just light waves that undergo interference as a result of the superposition of two waves. You can observe the same effect in microwaves, radio waves, sound waves and water waves.

You see interference effects when waves come together with a definite phase difference. For this reason, you don't see steady interference patterns on the surface of a choppy sea because the phase differences between the waves are continually changing. Waves only show stable interference effects when they are **coherent**. Light from a laser is coherent over quite long distances, and so easily produces interference.

Using white light, which contains a range of wavelengths, you can only see interference patterns when the path differences are small. This is why thick window panes don't show interference colours. For a soap film, the angles at which one colour is removed are very different from those at which a different colour is removed. But for a window pane, the places where different colours are removed all overlap.

Standing waves make music

Take a moment and hum the first few notes of your favourite tune. Imagine it being performed. If you are imagining a lead guitar solo you will be thinking of a very different sound to, say, a flute solo; these instruments would sound different even if they were playing the same tune. Not only can the principle of superposition explain why different instruments sound different, it can also explain why instruments can make a tune at all.

The sequence of notes is created on a musical instrument by way of **standing waves**. You may have seen standing waves on a slinky spring or perhaps on a vibrating cord illuminated with a stroboscope. You get standing waves when waves travel in opposite directions through one another. The standing waves on the strings of guitars and in the pipes of trumpets and clarinets are all examples of superposition: ones that make music.

When you pluck the string of a guitar, waves travel to each end of the string and bounce back and forth. Waves going in opposite directions produce a standing wave on the string, and the

Two waves will only show stable interference effects if they have a constant unchanging phase difference. If so they are said to be **coherent**.

coherent waves with constant phase difference

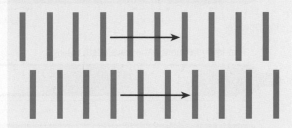

Atoms emit bursts of light waves. A burst from one atom is not in phase with a burst from another. So light waves from atoms are coherent only over quite short distances. This is why thin films give colours but thick window glass does not.

incoherent wave bursts with changing phase difference

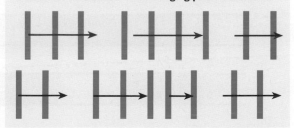

Coherent waves show stable interference effects

Strings of a guitar

at rest

fundamental $n = 1$

harmonic $n = 2$

harmonic $n = 3$

Key summary: standing waves in pipes

Closed pipes

loudspeaker

At a lower frequency, the wavelength is longer

A loudspeaker sends a sound into a long tube. Dust in the tube can show nodes and antinodes. Nodes are half a wavelength apart. So are antinodes. Dark colour shows maximum pressure variation and minimum motion of air (pressure antinode). Light colour shows minimum pressure variation and maximum motion of air (pressure node).

The fundamental: The lowest frequency which can form a standing wave has wavelength equal to twice the length of the tube.

Pipes open at both ends

Sound can be reflected from an open end as well as from a closed end.

This is how open organ pipes and flutes work.

Pipes closed at one end

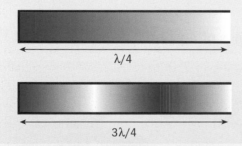

Pipes closed at one end are shorter, for the same note.

A clarinet is like this. An oboe is too, but with a tapered tube.

Some organ pipes are stopped at one end.

Frequencies of standing waves

	pipes open or closed at both ends strings fixed at both ends	pipes open at one end
length L	$L = n\lambda/2$	$L = (2n-1)\,\lambda/4$
fundamental	$f = v/2L$	$f = v/4L$
harmonics	$2f$	$3f$
	$3f$	$5f$
	...	...
	nf	$(2n-1)f$

The frequencies sounded by a pipe depend on the pipe's length

What is the source of vibrations for each instrument? How is the length of the standing wave changed to produce different notes?

string vibrates in one or more characteristic patterns. The vibrating string sets the body of the guitar vibrating, and the large surface area of the body efficiently transmits the sound to the air, so that an audience can easily hear it.

If the string is shortened by holding it down against a fret, the wavelength of the standing wave is shortened and the note produced goes to a higher frequency.

The note you hear is the frequency of the longest wave that will fit on the string, whose wavelength is twice the length of the string. This is called the **fundamental frequency**. But the quality of the guitar sound is produced by this sound being mixed with higher frequencies from shorter-wavelength standing waves called **harmonics** (chapter 3, p67).

Standing waves can also be formed in air. They make the notes in wind instruments. When you blow into a recorder the air passing the thin lip (fipple) below the mouthpiece forms rapidly spinning eddies. This sends a train of pressure waves down the hollow bore of the recorder, which reflect from the open end at the bottom of the instrument. As with the guitar, waves travel up and down the instrument and make a standing wave inside. The standing wave picks out one particular note, depending on the length of the open bore, usually the fundamental. If the player blows harder, higher frequencies are present in the eddies

around the lip, and a new standing wave of shorter wavelength can be picked out. Uncovering holes along the recorder effectively changes the length of the tube and so changes the wavelength of the standing wave. Once again, if the standing wave gets shorter, the frequency of the note rises.

The particular quality of sound of an instrument depends on the precise mixture of harmonics in the note, which the player can modify a little. It also depends on other things, such as how long the note lasts and the way it starts and stops.

Other kinds of standing wave include those in rod-like aerials to receive television signals (chapter 3, p65). People living beside inland lakes have puzzled about how sometimes the water can rhythmically rise and fall – the effect is called a *seiche* by French speakers living near Lac Leman in Switzerland. This movement is produced by the water in the lake slopping forwards and backwards in a huge standing wave on the two-dimensional lake surface. You will have seen a *seiche* if you have ever had a spill when carrying a shallow dish filled with water across to the sink. The quality of a human voice also depends on standing waves. The larger size of the male larynx explains why male voices are generally deeper than those of women. The folds of tissue (vocal cords) inside the larynx are larger and thicker in men and vibrate at lower frequencies. Their length and tension are controlled by many tiny muscles.

Quick check

1. A source of sound waves of frequency 570 Hz emits a note of wavelength 0.6 m in air at 20 °C. Show that the speed of sound at this temperature is about 340 m s^{-1}.

2. A stationary (or standing) wave is formed in a string with antinodes every 150 mm. Show that the wavelength of the standing wave is 300 mm.

3. A loudspeaker points directly at a wall 3 m away and emits a note of frequency 680 Hz. A standing wave is formed. If the speed of sound is 340 m s^{-1} show that the separation between minimum intensities (i.e. nodes) is 0.25 m.

4. A steady note of frequency 850 Hz from a factory siren is heard by two people standing in an open field. One is 20 m closer to the siren than the other. If the speed of sound in air is 340 m s^{-1} show that the path difference between the waves reaching the two people is 20 m and that this corresponds to 50 wavelengths.

5. Organ pipes can produce very low frequency notes. Show that an organ pipe closed at one end would have to be 2.8 m long to emit a note of 30 Hz (speed of sound = 340 m s^{-1}).

Links to the *Advancing Physics* CD-ROM

Practise with these questions:

30S Short answer *'Lloyd's mirror' for microwaves*

40S Short answer *Superposition of sound waves*

50S Short answer *Interference of sound waves*

90S Short answer *Stationary waves in a string*

100S Short answer *Standing waves in pipes*

Try out these activities:

50P Presentation *Slinky demonstrations*

60P Presentation *Superposition of microwaves*

110P Presentation *Standing waves in sound*

130P Presentation *Standing waves with microwaves*

Look up these key terms in the A–Z:
Amplitude, frequency, wavelength and wave speed; coherence; interference; phase difference; superposition; travelling and standing waves

Go further for interest by looking at:

10T Text to read *The Nautilus loudspeaker*

20T Text to read *Acoustics of rooms*

Revise using the revision checklist and:

20 OHT *Phase and angle*

40 OHT *Oscillations in phase*

60 OHT *Oscillations in antiphase*

80 OHT *Oscillations with a 90° phase difference*

6.2 What is light?

So far in this chapter we have described light as a wave. But as it turns out, the questions "waves of what?" and "waves in what?" lead into deep and interesting issues that physicists have had to face, including, surprisingly, questions about the nature of space and time.

Early ideas about light

Go back 2000 years. How would you have explained the iridescent shimmer of a butterfly wing or a beetle's back? Or any colour at all? Or even light itself? Light, though everywhere around us, has always been a mystery. Over the millennia some of the greatest names in philosophy and science, including Aristotle, Galileo, Descartes, Huygens and Newton, and writers including Goethe, have had their say in trying to account for it.

Some of the earliest ideas about light go back to Greek philosopher Leucippus of Miletus who lived around 450 BC. He speculated that maybe we see objects because their shapes travel across to our eyes. The idea was that very fine layers, like layers of skin off an onion, were continually peeling off and travelling away from things. Philosophers called these imagined layers eidola; later, the Romans called them simulacra. Needless to say, the notion raised more questions than it answered. What were eidola made of, and how did they get into the eye? The idea may seem very strange, but at least Leucippus had it the right way round; many others thought of seeing as if it were like touching, supposing that it is done by "visual rays" coming from the eyes, and returning to them from objects. They thought that an animal's eyes gleaming in the dark might be a glimpse of such rays. And Leucippus was facing an important mystery not fully resolved even today: how do objects "out there" in the world somehow "get into our minds" when we look at them? Or do we deceive ourselves into thinking that they do?

Ole Römer and the speed of light

In 1675 Danish astronomer Ole Römer (1644–1710), working at the Royal Observatory in Paris, became puzzled by the strange motion of the moons of Jupiter. He found that they kept a slowly varying timetable as they went round the planet. Over about six months they would gradually fall further and further behind schedule, and then in the rest of the year catch up again.

There had to be an explanation, and Römer thought of it. He realised that the moons could seem to get behind schedule if light from them took a finite time to reach the Earth from Jupiter. Then because the distance the light had to travel increased day by day as the Earth in its orbit round the Sun got farther and farther from Jupiter, the moons would seem to show up later and later. When the Earth started to get closer to Jupiter again, the moons would seem to catch up again on their timetable.

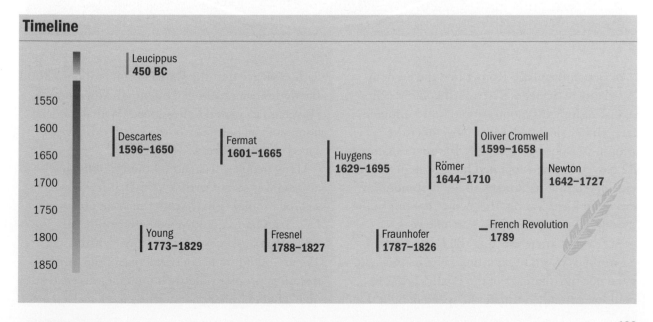

Timeline

Römer observes the moons of Jupiter get behind schedule

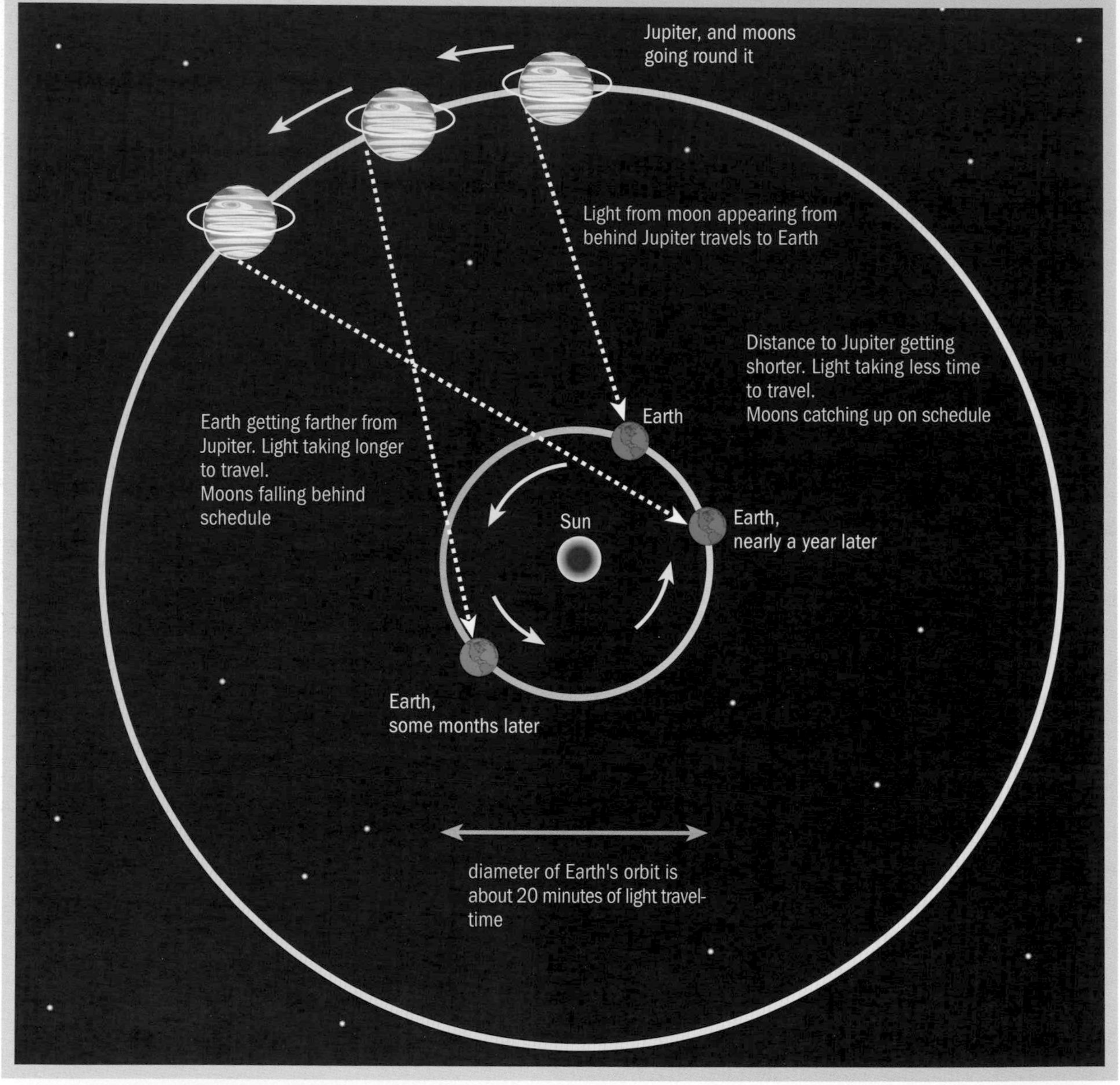

Jupiter, and moons
going round it

Light from moon appearing from
behind Jupiter travels to Earth

Distance to Jupiter getting
shorter. Light taking less time
to travel.
Moons catching up on schedule

Earth

Earth getting farther from
Jupiter. Light taking longer
to travel.
Moons falling behind
schedule

Sun

Earth,
nearly a year later

Earth,
some months later

diameter of Earth's orbit is
about 20 minutes of light travel-
time

In September 1676 Römer told the French Academy of Sciences that a forthcoming eclipse of one of Jupiter's moons would be 10 minutes later than calculated. On 9 November the eclipse was observed exactly 10 minutes late. On 22 November he explained his idea in detail and calculated that light takes about 22 minutes to cross the Earth's orbit, this being the shift in the moons' seeming timetable over one year.

All Römer knew was that the Earth's orbit is 22 minutes worth of light-time. Not having a good value for the diameter of the Earth's orbit, he couldn't actually estimate the speed of light. Later,

with evidence that the Earth's orbit is about 22 000 Earth diameters across, Dutchman Christiaan Huygens calculated light's speed. With a modern figure for the diameter of the Earth this gives a speed of 280 000 km s^{-1}.

Many people shrugged off these results; such a speed was so great, they said, that it might as well be infinity. That's what the great philosopher Descartes had said 30 years before, after all.

But Römer's work wasn't so easily dismissed. Twenty-odd minutes to travel one of the biggest distances people could then imagine was pretty quick, but definitely not nothing at all as Descartes' opinion required it to be. And the result raised new questions. Why that particular speed? What might light be made of that gives it this speed? As it happens, these problems lead straight into the deepest questions of modern physics and they also highlight the importance of a deceptively simple question: what is light?

Newton imagines particles

Isaac Newton (1642–1727) began his investigations into the nature of light in 1664 after buying a prism from a trinket stall at a fair. Newton finally published all of his work on light in 1704, in his book *Opticks*. The combination of mathematics and experiment found in *Opticks* became a model for future generations of investigators and can be instantly recognised as scientific to this day, even though people no longer believe all of its conclusions. *Opticks* is famous for the words, "Are not the rays of light very small bodies emitted from shining substances?" with which Newton gave support to a moving-particle picture of light.

But a reader of *Opticks* soon finds that Newton mixes together particle ideas with something very like wave ideas. He clings to the idea of particles, but suggests that they act by setting up vibrations in matter: "Do not several sorts of rays make vibrations of several bignesses, which according to their bigness excite sensations of several colours, much after the manner that the vibrations of the air, according to their several bignesses excite sensations of several sounds?" Remember that in Newton's time the word "several" meant "different". For "bigness" read "wavelength", and you are pretty much in a world of waves.

Newton knew of many wave phenomena, including colours in thin films – in fact he invented a way (Newton's rings) of measuring them accurately. He asks what might transmit vibrations from the Sun to the Earth as fast as Römer had shown they must travel. When he tries to account for what we now call interference effects, he says that particles of light must have, "alternate fits of easy transmission and easy reflexion", which is close to saying that like waves they have phases. However, on the particle side of the argument, he saw that refraction in glass could only be explained if the particles were pulled towards the glass

Isaac Newton began his investigations into the nature of light in 1664 after buying a prism from a trinket stall at a fair.

The highest praise: Einstein on Newton

"Nature was to him an open book, whose letters he could read without effort...In one person he combined the experimenter, the theorist, the mechanic and, not least, the artist in exposition. He stands before us strong, certain and alone: his joy in creation and his minute precision are evident in every word and every figure."

A modest man: Newton on Newton

"I do not know what I may appear to the world; but to myself I seem to have been only like a boy playing on the seashore and diverting myself in now and then finding a smoother pebble or a prettier shell than ordinary, whilst the great ocean of truth lay all undiscovered before me."

Christiaan Huygens (1629–1695) was born in Holland to a family of diplomats. He did work in France at the Royal court. His book *Traite de la Lumière* is addressed, in the introduction, to the French king.

Huygens drew this illustration of the "ether". He pointed out that ball A and ball D could both hit the row of balls, and that the movement of either one, passed through balls B and C, is rapidly transmitted to the other.

This image shows the title page of Huygens' *Traite de la Lumière*.

and speeded up inside it. It was not until 1850, a century and a half later, that Jean Foucault and Armand Fizeau both managed to show directly that light actually goes slower in water than in air, (chapter 1). Newton had it the wrong way round.

Newton did not come down conclusively on one side or the other, but popular explanations such as the curiously titled *Newtonianismo per le dame* (*Newtonianism for the ladies*), written by Francesco Algarotti in 1735, decided that he had. They told the world that Newton believed in a travelling particle picture, and long after his death it was widely assumed to have his backing. In retrospect, "the dead hand of Newton" may well have hindered the development of alternative pictures of light.

Huygens imagines waves

Christiaan Huygens (1629–1695) was an independent-minded and prolific Dutchman born into a well educated family of diplomats. Not only did he deduce a value for the speed of light from Römer's observations, he also made new observations on Saturn's ring system, discovered Saturn's largest moon Titan, studied the mechanics of pendulums and invented accurate clocks.

It was from Huygens that, two years after Römer's announcement about Jupiter's wayward moons, the French Academy of Sciences again heard a revolutionary development in thinking. Huygens stood before them to present his ideas on the wave nature of light. Römer was present and heard his ideas of the finite speed of light used to introduce Huygens' wave concept.

Huygens argued from commonplace observation. He noted the "obvious" fact that light doesn't get in the way of light; that two torches each illuminate the other as if the other were not shining. He pointed out that travelling light particles would be expected to hit one another, so that the light from one torch should collide with the light from the other. He even remarked that this made it hard to see how two people could look one another in the eye at the same moment. The observation that water waves pass through each other without hindrance – or in modern terms, that they superpose – led Huygens to picture light as a wave motion.

This picture left some difficult questions to answer: "If light is a wave then what is it a wave in?" and "How can the wave picture explain the fact that light travels in straight lines and is reflected and refracted?"

Ethereal matter

Huygens thought of light as a longitudinal wave, like sound. He drew pictures of how the wave must travel: very tiny, very hard particles in contact that pass the wave motion from one to the next. Using an analogy, Huygens described how this works with spheres arranged in a line (see image, opposite): "One finds, on striking with a similar sphere against the first of the spheres, that the motion passes as in an instant to the last of them…and one sees that the movement passes with an extreme velocity which is the greater, the greater the hardness of the spheres."

You can probably see where his argument is going. The velocity of light had just been measured and light went far, far faster than anything else. Therefore, what it goes through (the "ether") must be composed of particles that are much harder than anything else. This gives him a problem. How do we all move through this mass of hard particles without noticing them? How do the planets go through them without slowing down?

The idea of space filled with a material substance to carry light waves has not survived, though it was a long time dying. But another of Huygens' ideas has proved tremendously fruitful: this is his idea of "wavelets" to explain how waves recreate themselves.

Huygens' wavelets

Dip a stick into a pool of water. A ripple spreads out evenly in all directions. Huygens imagined light spreading out in a similar manner: "[light] spreads, as sound does, by spherical surfaces and waves: for I call them waves from their resemblance to those which are seen to be formed in water when a stone is thrown into it, and which present a successive spreading as circles, though these arise from another cause and are only in a flat surface."

Waves, according to Huygens, come from every part (see A, B, C on image, right) of a candle flame. A little distance from the candle they all seem like spheres centred on the candle.

Huygens had a simple idea about how waves get from place to place, changing form as they go. The idea was that every point on a wave front acts in the same way as the circular ripples sent out in every direction produced by a stone dropped in water. The new wave front is just the places where the wavelets are in phase with one another, and so combine to recreate the wave. Everywhere else the wavelets from one part of the wave are out of phase with those from another and the result is nothing.

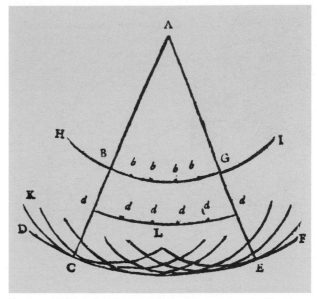

Huygens tried to show how a spherical wave expands. His idea is that the wave front DCEF is made of wavelets from the earlier wave front HBGI, starting from all the places like b,b,b,b on that wave front.

Waves, according to Huygens, come from every part (A, B, C) of a candle flame. A little distance from the candle they all seem like spheres centred on the candle.

The wavelet idea is more radical than it looks. It denies that waves simply roll grandly forward just as if they know where to go, even if that's how it looks. It says that you have to think of wavelets from the wave as spreading everywhere, but adding up to nothing everywhere except where the wave actually goes. Waves propagate, according to Huygens, because of superposition of wavelets. A similar idea has surfaced again in modern

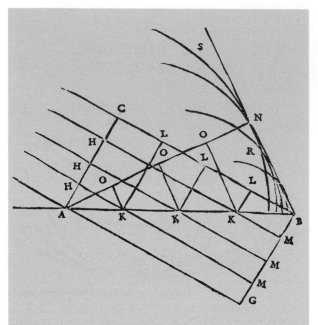

Huygens explains how a wave is reflected. The incoming (plane) wave is CA. From A a wavelet expands, becoming the circle SN centred on A. Parts of the wave front CA labelled H,H,H arrive later at the mirror, at points K,K,K, and produce wavelets such as R and L. All these wavelets combine to produce the reflected plane wave BN. The pure geometry ensures that the angles of incidence and reflection are equal.

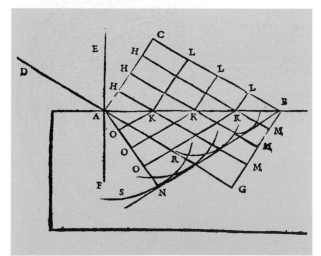

Huygens explains refraction. The incoming wave is CA. Wavelets start out into the glass from points such as K,K,K. They go more slowly in glass than in air (points O,O,O in glass are closer than points L,L,L in air), so the wave front BN in the glass is swung downwards on the diagram.

quantum physics (chapter 7), and has further changed ideas about what light is.

Huygens did something else impressive. He proved that on his wave theory the bent path of refracted light is the path that takes the shortest time, even though longer in distance than the direct straight path. Thirty years earlier, Fermat – a mathematician – had the same idea but lacked a theory of light to account for it.

You may have been struck by how the ideas of Newton, Huygens and others depend on simple analogies with well known things. Newton thought that light was like travelling particles, whereas Huygens believed it was like a travelling wave of sound. You might ask yourself whether this is scientific. If you think not, then ask where else can people start except with what they know?

What did Huygens achieve?

Huygens imagined that wavelets made it possible to see how a wave propagates, how light meeting a mirror obeys the law of reflection:

angle of incidence = angle of reflection

and how a beam of light is refracted, that is, how it bends as it passes from air into water or glass.

He also made a prediction: that the speed of light should be slower in water and glass than in air. Nobody could test this at the time, but in the end it turned out to be right. The speed measured by Fizeau and Foucault (p136) did change in the right way and by the right amount. All explained just by superposing wavelets.

But Huygens' idea also posed a huge problem. What could these waves be travelling in? Huygens called it ether, but this just gave the problem a name, not a solution. What could the ether be like? How could the stars and planets slip so easily through such a substance?

Sir George Stokes (1819–1903) suggested that the ether was like a thick mixture of glue and water that acts like a solid for rapid vibrations but allows slow movement with relative ease.

You can make your own Stokesian ether by adding a small amount of water to a few tablespoons of custard powder or cornflour to make a very thick cream. This non-Newtonian fluid will allow you to move a spoon through it slowly but will resist quicker movement. If you get the mixture just right you can shatter it with a hammer! Soon, however, electromagnetic theory brought about the end of such mechanical models of the ether (p145).

Quick check

1. Römer interpreted his findings to suggest that light takes about 20 minutes to cross the diameter of the Earth's orbit. Using the modern value for the radius of the Earth's orbit, 1.5×10^8 km, show that this gives a value of $250\,000$ km s^{-1} for the speed of light. Was Römer's estimate of the time too large or too small?

2. Why, knowing that light going from air to glass is refracted towards the normal, would Newton have supposed that particles of light must travel faster in glass than in air?

3. Suggest one argument in favour of picturing light as a stream of particles, and one against.

4. Suggest one argument in favour of picturing light as a wave, and one against.

5. By 1973 the internationally recommended value for the speed of light was $c = 299\,792\,458 \pm 1$ m s^{-1}. Show that this represents a fractional uncertainty of 3 parts per billion.

6. The first reliable measurement of the speed of light done on the Earth's surface, by Fizeau in 1849, timed flashes of light out and back over the 8.6 km distance from Montmartre to Suresnes. Show that the time interval Fizeau had to be able to measure was about 60 μs.

Links to the *Advancing Physics* CD-ROM

Practise with these questions:
110S   Short answer *Measuring the speed of light*
250S   Short answer *Specifications for a lens*
120X   Explanation–exposition *Checking out a mirror*

Try out these activities:
40S   Computer screen *Overlapping ripples*
50S   Computer screen *Ripple tank images*
180S   Software-based *Designing parabolic mirrors*
200S   Software-based *Trip times for a lens*

Look up these key terms in the A–Z:
Huygens' wavelets; reflection; refraction; speed of light

Go further for interest by looking at:
30T   Text to read *Historical attempts to measure the speed of light*

Revise using the revision checklist and:
190   OHT *An estimate of the speed of light*
200   OHT *Huygens' candle*

6.3 Wave behaviour understood in detail

Thomas Young revives wave theory

In 1801, at the Royal Society in London, a medical doctor gave an important physics lecture. Today you would expect a physicist to give the talk and the doctor to stick to medicine, but two centuries ago, people really admired men and women who knew a bit (often a great deal) about many subjects. Beside medicine, the doctor on the podium that day, Thomas Young (1773–1829), knew a dozen languages, wrote poems, did physics experiments and helped to decipher Egyptian hieroglyphics on the Rosetta Stone. But by all accounts he wasn't at all good at explaining himself clearly, which made it hard for him to get his ideas respected.

The subject of his lecture was the wave theory of light. In England, most people, partly out of respect for Newton, held to particle theory. Young wanted to show that wave theory was alive and kicking. He reminded his audience of Newton's many observations of what look like wave phenomena – colours in soap films and bright and dark fringes.

Young went on giving such lectures. One of his most convincing demonstrations was the very simple double-slit experiment that, along with the modulus of elasticity (chapter 4), now bears his name. Young shone light through two barely separated pinholes onto a distant screen. On the screen appeared a regular pattern of bright and dark fringes. Off-centre, the light from one pinhole had farther to go than the light from the other, and the two could get out of step. Where they did, the screen would be dark. A little farther off-centre the two would be in step again, and the screen would be bright. The most convincing part was if one pinhole was covered up: the bright and dark fringes vanished.

Looking at the fringes, you are in a way looking at the tiny light wavelength blown up in scale. The spacing x between the fringes is the wavelength λ multiplied by $\frac{L}{d}$ (which is at least 1000 and possibly 10000, since the slit to screen distance L is more than a metre and the slit spacing d is less than a millimetre). That is:

$$x = \lambda \frac{L}{d}$$

Young expressed the idea of wave superposition

Thomas Young (1773–1829) was a polymath – linguist, archaeologist, poet and medical doctor. He lectured badly, it seems, and scornful reviews of his presentation of the wave theory of light are to be found in journals of the time.

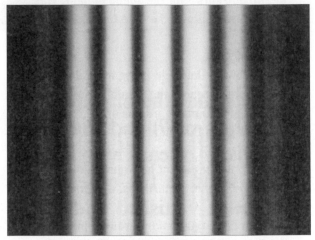

An example of the fringe pattern obtained from two slits.

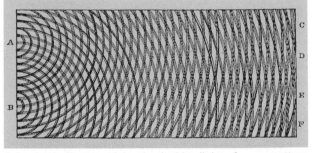

A drawing by Young explains the double-slit interference pattern (from *Lectures on Natural Philosophy 1807*). Look along the page from the left and you can see curved lines where peaks and troughs coincide, their ends are marked C,D,E,F.

as, "When two undulations, from different origins, coincide either perfectly or very nearly in direction, their joint effect is a combination of the motions belonging to each." To understand Young in modern terms, for "undulations" read "waves", for "combination of the motions" read "superposition".

Young's audiences were sometimes far from impressed. A reviewer wrote savagely of, "…the feeble lucubrations of this author, in which we have searched without success for some traces of learning, acuteness and ingenuity, that might compensate his evident deficiency in the powers of solid thinking, calm and patient investigation and successful development of the laws of nature, by

Key summary: Young's double-slit interference experiment

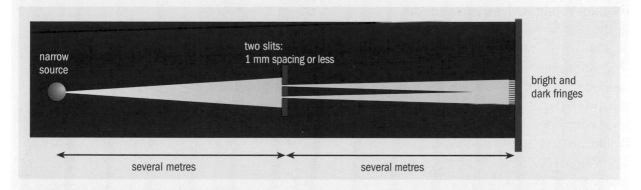

narrow
source

two slits:
1 mm spacing or less

bright and
dark fringes

several metres   several metres

Geometry

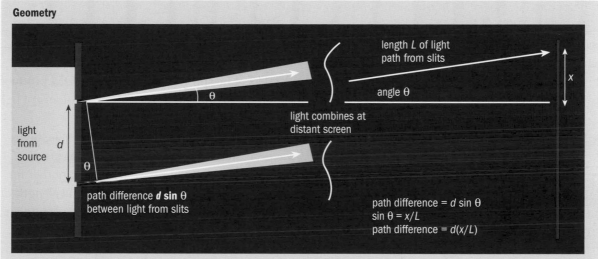

length L of light
path from slits

angle θ

x

light combines at
distant screen

light
from
source

d

θ

path difference $d \sin \theta$
between light from slits

path difference = $d \sin \theta$
$\sin \theta = x/L$
path difference = $d(x/L)$

Approximations: angle θ very small; paths effectively parallel; distance L equal to slit–screen distance.
Error less than 1 in 1000

Two simple cases

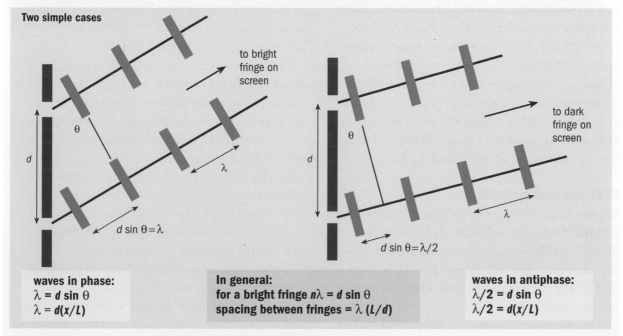

to bright
fringe on
screen

θ

d

λ

$d \sin \theta = \lambda$

to dark
fringe on
screen

θ

d

λ

$d \sin \theta = \lambda/2$

waves in phase:
$\lambda = d \sin \theta$
$\lambda = d(x/L)$

In general:
for a bright fringe $n\lambda = d \sin \theta$
spacing between fringes = $\lambda (L/d)$

waves in antiphase:
$\lambda/2 = d \sin \theta$
$\lambda/2 = d(x/L)$

Wavelength can be measured from the fringe spacing

This micrograph shows the surface of a diffraction grating where thousands of fine lines are ruled very close together on glass.

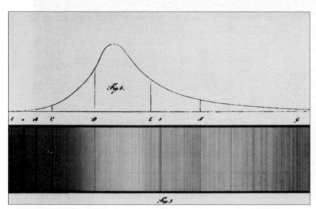

A portion of the solar spectrum drawn by Joseph Fraunhofer.

steady and modest observation of her operations".

It was a youthful French engineer Auguste Fresnel (1788–1827) who really changed scientific opinion, through brilliance and hard work. His prize-winning book of memoirs of 1819 reports accurate experiments, explaining each experiment in detail from the wave point of view, as exactly as possible. The mathematical ideas that he developed are still used today.

Gratings and spectra

We began this chapter by talking about the blue wing of the *Morpho* butterfly (p123). The many tiny platelets on the wing, arranged in regular rows, sharply reflect just one colour. It is the regular rows that do the trick, and the same idea was used by Joseph Fraunhofer in 1821 when he became the first to obtain spectra using a **diffraction grating**. Thousands of fine lines ruled very close together on glass or on a mirror replaced the rows of

platelets. The spiral track on a CD-ROM produces the same effect, seen in the rainbow colours of light reflected from it.

With the equivalent of thousands of slits, light waves scattered from each line can – at just the right angle – all be in phase with one another for a particular wavelength. This is the wavelength of the colour you see. As a new scientific instrument the diffraction grating proved immensely useful. Indeed it made it possible for the first time to know what the stars are made of: the elements they contain show up with their spectral fingerprints in the light from the star. Famously, the element helium was discovered in the spectrum of the Sun before being isolated in the laboratory.

The grating is now an essential tool of modern physics, pure and applied. High-precision gratings carefully and accurately ruled on metal and glass are expensive to make, but cheap multiple plastic copies made from a single original are good enough for many purposes.

Spectra taken with gratings can be obtained for infrared and ultraviolet light as well as for visible light. These often require reflection gratings because a transmission grating would absorb too much of the radiation passing through it.

At the beginning of the 20th century, regular rows of atoms in materials started to be used as gratings for the much smaller wavelength X-rays, first to show that X-rays have wavelike behaviour, but later as a tool to investigate the arrangements of the atoms themselves inside materials.

Key summary: diffraction grating

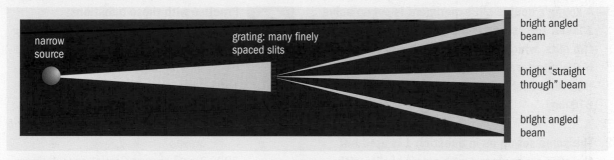

Geometry

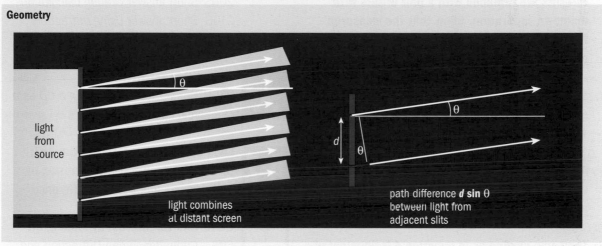

Waves from many sources all in phase

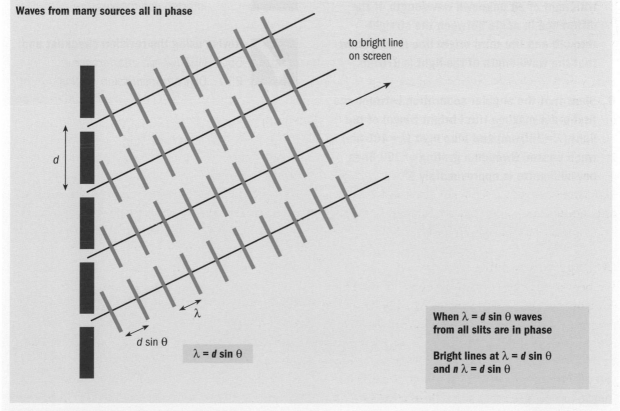

$$\lambda = d \sin \theta$$

When $\lambda = d \sin \theta$ waves from all slits are in phase

Bright lines at $\lambda = d \sin \theta$ and $n \lambda = d \sin \theta$

Sharp bright spectral lines at angles where $n \lambda = d \sin \theta$

Quick check

1. A Young's double-slit experiment is set up using a laser. The screen is placed 3 m away from the two slits, which have a separation of 0.4 mm. If the interference fringe spacing is 5 mm show that the wavelength of the laser light was 670 nm.

2. The same set-up as in question 1 is used but the slits are replaced with a pair of slits with unknown separation. Show that the measured fringe spacing of 2 mm implies a slit separation of 1 mm.

3. Show that the slit spacing of a diffraction grating marked as 80 lines per mm is 1.25×10^{-5} m.

4. A grating with spacing 1×10^{-5} m is illuminated with light of wavelength 550 nm. Show that the angle of the first-order bright line is about 3°.

5. The same grating as in question 4 is illuminated with light of an unknown wavelength. If the difference in angle between the straight-through and the third bright line is 10°, show that the wavelength of the light is 870 nm.

6. Show that the angular separation between the first-order maxima (first bright fringe) of red light ($\lambda = 700$ nm) and blue light ($\lambda = 400$ nm) when passed through a grating of 300 lines per millimetre is approximately 5°.

Links to the *Advancing Physics* CD-ROM

Practise with these questions:
130X Explanation–exposition *Calculating wavelength in two-slit interference*
150S Short answer *Questions on the two-slit experiment*
190S Short answer *Superposition of radio waves*
200S Short answer *Grating calculations*
210S Short answer *Using diffraction gratings*

Try out this activity:
250H Home experiment *Using a CD as a reflection grating*

Look up these key terms in the A–Z:
Diffraction; double-slit interference; gratings and spectra; path difference

Go further for interest by looking at:
60S Computer screen *Diffraction and interference for pleasure*

Revise using the revision checklist and:
550 OHT *Two-slit interference*
650 OHT *A transmission grating*

6.4 Looking forward

The wave picture of light went from strength to strength, except for the awkward fact that nobody knew what light was waves of. The next big step was taken in 1865 by James Clerk Maxwell, who followed up a speculative idea put forward 20 years before by Michael Faraday.

Faraday had written in an essay entitled "Thoughts on ray vibrations": "The view which I am so bold as to put forth considers, therefore, radiation as a high (frequency) species of vibration in the lines of force…It endeavours to dismiss the ether, but not the vibration."

By "lines of force" Faraday means a field, as shown by the pattern of iron filings around a magnet. No more material ether; no imaginary wave-transmitting substance indetectably filling empty space; just vibration. Maxwell turned this visionary idea into what is now called the electromagnetic theory of light. And as you know, modern communications have sprung from that idea (chapter 3).

The next chapter reveals that the story is not yet over. Quantum physics takes a hint from Faraday, who wanted to throw away the ether and keep just the vibration, but goes further still. The idea is to strip away all the stories about waves in this or in that, and keep just "how to work things out". The rotating arrows called phasors, introduced at the start of this chapter, are just the tool that is needed.

Waves spread out from apertures

Both Young and Newton knew that Italian Francesco Grimaldi (1618–1663) had long before shown that if you put an obstacle in front of a small source of light then the light bends into the shadow, with coloured fringes around the shadow of the edges. This is diffraction.

Fresnel put Huygens' wavelet idea to careful mathematical use and gave a precise explanation of the patterns of light and dark in convincing quantitative detail. This makes it possible to predict, for example, by how much the beam of a microwave transmitter from a dish aerial will spread.

Huygens' wavelet idea says, "imagine every part of the aperture (radar dish, or slit through which light passes) as sending out wavelets. Then for a given place receiving the waves, just add up all

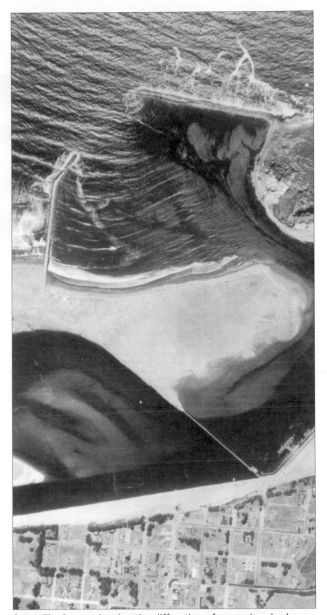

A satellite image showing the diffraction of waves in a harbour.

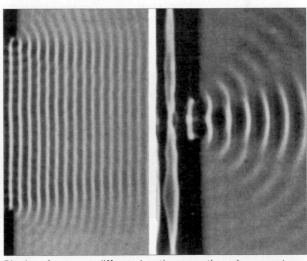

Ripples of water are diffracted as they pass through an aperture.

Key summary: diffraction at a single aperture

Single slit

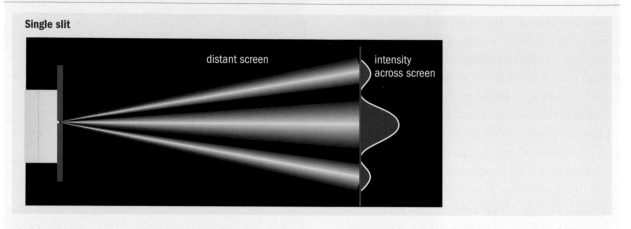

distant screen

intensity across screen

Simplified case: screen distant, paths nearly parallel

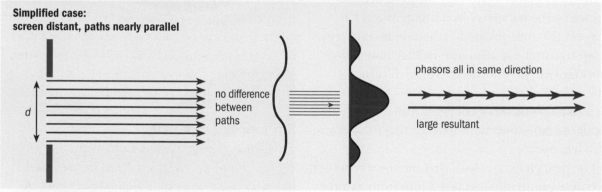

d

no difference between paths

phasors all in same direction

large resultant

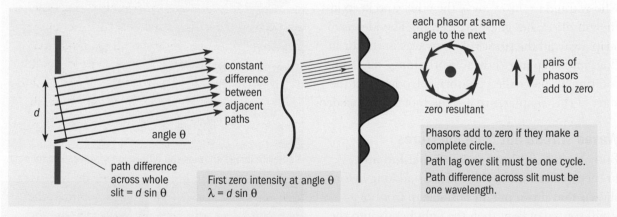

d

constant difference between adjacent paths

angle θ

path difference across whole slit = $d \sin \theta$

First zero intensity at angle θ
$\lambda = d \sin \theta$

each phasor at same angle to the next

pairs of phasors add to zero

zero resultant

Phasors add to zero if they make a complete circle.
Path lag over slit must be one cycle.
Path difference across slit must be one wavelength.

Useful approximation

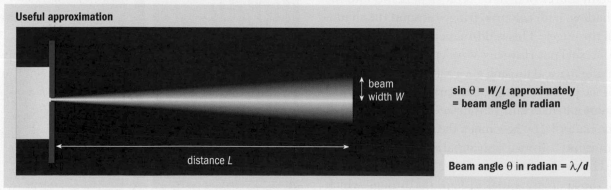

beam width W

distance L

$\sin \theta = W/L$ approximately
= beam angle in radian

Beam angle θ in radian = λ / d

To a good approximation, a beam passing through a slit of width d spreads into an angle $\theta = \dfrac{\lambda}{d}$

of the wavelets." The phasor picture makes this recipe easy to follow. Every wavelet has a phasor that rotates at the wave frequency as it travels, going once round for every cycle and so once round for every wavelength of path travelled. To get the total wave amplitude at any particular place you just add up the phasors for all the wavelets arriving at that place.

Going to a place straight ahead of the aperture and a long way away, all the wavelets have travelled the same distance and so have the same phase. Their phasor arrows point in the same direction and they add up to a large amplitude.

Going to a place off to one side, wavelets from one edge of the aperture have farther to go than ones from the other side. So a phase lag builds up between wavelets coming from successive parts of the aperture. Now the phasors, when added together tip to tail, tend to curl up. When they curl up into a complete circle, the last phasor meets the tail of the first phasor and all the wavelets arriving at that point on the screen add up to nothing. For this to happen the total phase lag across the whole aperture must be one cycle or, in terms of path difference, one wavelength. The result is that the edge of the beam where the waves first have zero intensity is at angle θ where:

$$\lambda = d \sin \theta$$

or, if the angle is small, where:

$$\text{beam angle } \theta \text{ in radians} = \frac{\lambda}{d}$$

Your eye has just such an aperture: the pupil. It is interesting to think how the pupil affects the resolution of your eye, and whether it is the size of the pupil or the micrometre spacing of rods and cones in the retina that most limits the resolution.

Resolution of optical instruments

You may have wondered why radio telescopes and radar signalling equipment are built with such large reflecting dishes. Whether the signal is being sent out or is being received through the aperture of a dish, the angle of the beam in radians is of the order of magnitude $\frac{\lambda}{d}$. (For a circular aperture in place of a narrow slit, a more exact value is $1.22\frac{\lambda}{d}$.)

This means that a radio telescope or radar aerial can't tell apart distant objects that are closer than this angular distance apart. The objects cannot be **resolved**. Modern radio astronomy solves the problem by using different telescopes as far apart as possible on the Earth, but linked to act as if they were a single virtual telescope.

Diffraction similarly limits the resolution of optical telescopes. The need for higher and higher resolution in astronomy, to see fine details of distant stars and galaxies, has led to telescope mirrors being made larger and larger. The mirror of the Hubble Space Telescope is 2.4 m in diameter – limited by what can be lifted into orbit – but ground-based telescopes can be several times larger than this. Besides high resolution, a larger telescope mirror also collects more light, making it possible to detect ever-fainter and more distant objects. In astronomy, big is better.

Stripped-down thinking

The phasor picture gives a rather simple description of diffraction; indeed Fresnel used much the same reasoning in analysing a whole range of diffraction and interference phenomena. The diagram in "Key summary: diffraction at a single aperture" (opposite) is peculiar in a significant way. It makes almost no mention of waves: just spinning phasor arrows. Superposition just becomes adding arrows tip to tail. Even the wavelength λ is just the distance over which a phasor arrow turns round once.

"So what?" you may ask. Surely the phasor arrows are designed to work as they do because light is a wave motion? In the next chapter, there is a surprise in store. We will, in describing the world of quantum physics, turn this argument back to front. Light, we shall say, exchanges energy in lumps called photons. Why then does light behave like a wave? The answer will be that light seems to behave as a wave because phasor arrows governing the photons behave as they do.

As sometimes happens in physics, what started as an abstract mathematical device (spinning phasor arrows) for performing calculations turns out to give a deeper view of reality, by unhooking the mind from one picture (waves) and opening up new ways of seeing things.

Try these

1. A 1 m diameter radar dish on a telecommunications tower transmits radar waves of wavelength 10 mm. Show that the beam width at a range of 10 km is about 100 m.

2. Sketch the phasor arrows for bright and dark fringes in a Young's double-slit experiment.

3. The pupil of your eye has a diameter of about 5 mm and the diameter of your eye is about 25 mm. The wavelength of light is of the order of 500 nm. Show that the limit on the angular resolution of your eye set by the size of the pupil is about 10^{-4} radian, and that this corresponds to an image about 2.5 μm wide on your retina.

4. The 300 m diameter radio-telescope dish at Arecibo in Puerto Rico is sunk in a shaped hollow in the ground. Suppose that unevenness in its surface makes 100 mm be the shortest wavelength that is usable. Show that the angular resolution at this wavelength is about 5×10^{-3} radian.

5. The amplitude of waves through a single narrow slit, in the straight through direction where the phasors line up in a straight line, is A. Show that an angle to the side where the phasors curl into a semicircle (so that the resultant is the diameter of the semicircle), the amplitude is $2A/\pi$.

6. Look at the ripples diffracted by the narrower slit in the photograph on p145. The angle between the direction of the first minimum and the straight through direction is about 30°. Show that the ratio of the width of the slit to the wavelength is about 2.

Links to the *Advancing Physics* CD-ROM

Practise with these questions:
220S Short answer *Phasors to oscillations*
230S Short answer *Wave to phasor*
240S Short answer *Linear and angular speed*

Try out these activities:
260S Software-based *Introducing phasors*
270S Software-based *Amplitude and frequency of oscillations with phasors*
280S Software-based *Two phasors at once*
320S Software-based *Young's slits by phasors*

Look up these key terms in the A–Z:
Diffraction; phase and phasors; superposition

Go further for interest by looking at:
50T   Text to read *Radio waves: fading and interference*

Revise using the revision checklist and:
70S   Computer screen *Modelling two slits with phasors*
80S   Computer screen *Calculating one phasor*
900   OHT *Diffraction*

Summary check-up

Superposition ✓

- Interference, diffraction and standing waves can be produced by the superposition of waves
- Two waves of the same frequency combine by adding their amplitudes if in phase, or by subtracting their amplitudes if in antiphase
- Coherence is necessary for a stable interference pattern
- Standing waves on strings or in pipes have a fundamental frequency, with harmonics that are whole-number multiples of the fundamental
- Wavelength can be calculated from Young's two-slit experiment using $\lambda = x\dfrac{d}{L}$ where x is the fringe spacing, d is the slit separation and L is the distance from the slit to the screen
- Wavelength can be calculated for diffraction gratings using the relationship $n\lambda = d\sin\theta$, where θ is the angular displacement of a bright fringe, n is the fringe order and d the slit separation
- Waves passing through an aperture of width d spread out, typically over an angle θ given by $\lambda = d\sin\theta$

The phasor picture ✓

- Phase differences can be measured as angles, using rotating "clock arrows" (phasors)
- Phasor arrows have length equal to the wave amplitude and rotate at the wave frequency
- Wave superposition is represented by adding phasor arrows tip to tail
- The phasor picture enables simple quantitative calculations to be done starting from Huygens' wavelet idea

The nature of light ✓

- The finite speed of light could be detected in delays in the regular motion of Jupiter's moons
- During the 18th century, a moving-particle model of light was widely held, and its supporters claimed the authority of Newton
- Huygens' wavelet idea explains wave propagation as the superposition of wavelets from every point on a wave front, so creating a new wave front. It explains reflection and refraction of light
- Wave theories of light such as Huygens' had the problem of trying to describe the ether in which the waves were supposed to propagate
- Young and Fresnel provided support for wave theory by calculating the shape of interference and diffraction fringes in a range of experiments
- Fraunhofer and others made precision observations of spectra using the diffraction grating as a tool

Questions

Useful data: Speed of light = $3 \times 10^8\,\text{m}\,\text{s}^{-1}$

1. **The back of a particular beetle looks bright blue from a certain angle, but not from other angles. Assume that this is due to diffraction by ridged layers of plates on the surface of the beetle's back, laid out like overlapping tiles.**

   (a) Make a sketch showing how bright blue might be scattered from successive ridges so as to be seen strongly reflected at a particular angle.

   (b) What is the order of magnitude of the spacing between ridges?

   (c) If the beetle were to evolve so as to look red at the same angle, how would the layout of the ridged "tiles" on its back have to change?

2. **A two-slit interference experiment is to be set up for an Open Day. It will be done with laser light, wavelength 600 nm. The closest the slits for the laser experiment can be ruled is 0.3 mm.**

   (a) How many wavelengths apart are the slits?

   (b) If the screen is set 3 m from the slits, will visitors be able to see the bright fringes distinctly? What will be their spacing?

   (c) A visitor calculates that the ratio of the fringe spacing to the distance from slits to screen is 1:500. Explain how this is connected to the slit spacing, measured in wavelengths.

   (d) The experiment takes up a lot of space. If the distance from slits to screen were reduced to 1 m, what effect would that have on the fringe spacing?

   (e) It is suggested that a second two-slit experiment should be set up alongside, using microwaves of wavelength 30 mm, as a scaled-up version of the laser experiment. A student objects that the two slits would need to be ridiculously far apart. Is the objection reasonable?

   (f) After some discussion, the slits for the microwave version are put 150 mm apart. Make a scale drawing to show the direction in which to expect the first bright fringe.

   (g) Very roughly, what would be the fringe spacing in the microwave experiment if the detector were set 1 m from the slits?

3. **The waves carrying a microwave communications link have a frequency of 15 GHz. They are made into a beam by a reflecting dish 2 m in diameter.**

   (a) At what rate does the phasor representing the wave rotate?

   (b) How far does the wave travel in one rotation of its phasor?

   (c) How is the answer to (b) related to the wavelength of the waves?

   (d) Draw a sketch using phasors to show how the beam, although it is strong along the axis of the dish, can fall to near zero at a small angle to the axis. (Treat the 2 m dish as a slit 2 m wide.)

   (e) Estimate the width of the beam from the dish at a distance of 1 km.

4. **Standing waves, interference, diffraction, colours in thin films, are all examples of wave superposition. Wave superposition is therefore involved in music and art as well as in scientific and technical work. Choose one example of an application of wave superposition effects in any field, and use it to explain:**

   (a) what the superposition effect is in your example, and why you count it as a case of superposition;

   (b) the use to which superposition is put in your example;

   (c) how to calculate any quantities relevant to your example;

   (d) reasons for choosing the frequencies or wavelengths of the waves in your example.

5. **Imagine that you have heard a lecture by Römer on his ideas about light. The audience is invited to raise any objections that they have.**

   (a) Remind the audience of one thing Römer is likely to have said, and state your objection, with a reason.

   (b) Suggest how a member of the audience might respond to a lecture given by either Huygens or Newton.

6. **The reed of a clarinet effectively closes it at one end, the other end of the pipe being open.**

   (a) Sketch the standing wave in the clarinet when it is playing the lowest note possible.

   (b) Sketch the standing waves for the first two harmonics above the fundamental.

   (c) Relate the length of a given type of clarinet to the range of notes it can play. Start from knowledge of, or an estimate of, either length or frequency, and relate it to the other.

7 Quantum behaviour

Fundamental particles of matter – electrons, quarks and neutrinos among others – act like nothing you have seen before. But they all share the same remarkable style of acting: quantum behaviour. Photons, the fundamental particles of light, do it too. In this chapter we will:

- explain why light needs to be thought of as photons
- use photons as a simple example of quantum behaviour
- show that electrons behave in that way too
- prepare the way for the modern picture of fundamental particles

7.1 Quantum behaviour

Quantum behaviour is important and fundamental. It accounts for the way that atoms

Dyson on Feynman

"...Dick Feynman told me about his...version of quantum mechanics. 'The electron does anything it likes,' he said. 'It just goes in any direction at any speed, forward or backward in time, however it likes, and then you add up...'

I said to him: 'You're crazy.' But he wasn't."

Freeman Dyson Some Strangeness in Proportion *(1980) ed. H Woolf (Addison Wesley) p376*

behave, and for the ways in which atoms combine to form molecules. It underlies how the most fundamental particles behave. Essential on these small scales, quantum behaviour shows up on large scales too – for example, in superconducting coils in magnets used in medical scanning.

Quantum things have their own way of behaving that is completely unlike the behaviour of everyday objects. The rule for quantum behaviour

Key summary: getting about

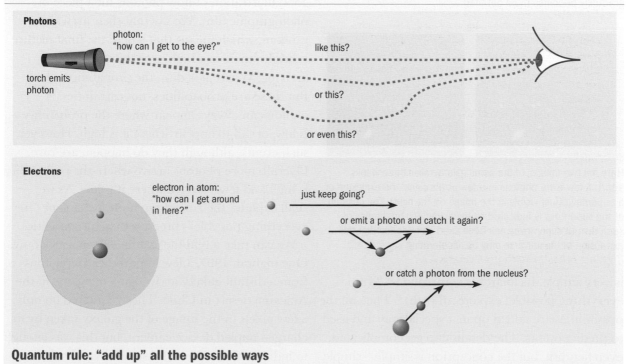

Quantum rule: "add up" all the possible ways

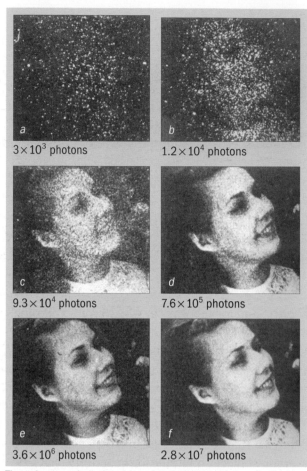

a 3×10^3 photons

b 1.2×10^4 photons

c 9.3×10^4 photons

d 7.6×10^5 photons

e 3.6×10^6 photons

f 2.8×10^7 photons

These images show the same picture taken with successively more and more photons. With very little light, the picture is a random blur of grains; as more photons arrive the picture gradually becomes more detailed.

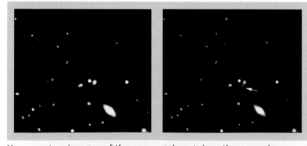

Here are two images of the same galaxy taken three weeks apart. A few extra photons in a few pixels signal the explosion of a supernova (just visible in the image on the right). The position of the supernova is indicated by the arrow. The brightness of such distant supernovae has been used as evidence that the expansion of the universe may be accelerating.

is very simple. Its anarchic command is: "Try everything possible; explore all paths." Then all the possibilities are added up in a special way and used to predict events. The details can get complicated, even fiendish, but the conception is simple – simple, grand and very general.

New ideas always seem peculiar at first, but you get used to them in time. That's how it is with quantum behaviour. Start young and you stop worrying sooner!

We will limit most of the story of quantum behaviour to the basic problem of how particles travel from one place to another. The same ideas work for how particles change into one another, or interact with one another. In other words, they work for all of quantum physics.

First we will tell the story of the quantum behaviour of photons, because it's the simplest one. Later we will show how electrons behave in essentially the same way, though this story is more complicated because of their mass and charge.

Light arrives randomly in lumps

Take photographs in progressively dimmer light. You might expect that less light will just give a fainter photograph, but it doesn't. Instead, the photograph breaks up into randomly arranged grains, just as if the light energy were arriving randomly in lumps. Turn up the brightness and the lumps of light arrive thick and fast, producing the smooth-looking picture you expect. The lumps, or **quanta**, of light are called **photons**.

The images in the box on the left are a good illustration of how photons hit a piece of photographic film. We say that their arrival is random, which means that where the final picture is bright, there is a high probability photons arriving. Where it is dim, the probability is low. But these are probabilities, not certainties. A photon can always appear where the probability is low, or fail to appear where it is high. However, such events, although they do happen, are rare. Overall, more photons arrive where the probability is high and fewer arrive where it is low. As we shall explain, the rule of quantum behaviour ("try everything possible") predicts these probabilities.

We can take a real-life example from astronomy. One night in 1997, a few hundred extra photons from a distant galaxy arrived at a telescope in the Atacama desert in Chile. They registered on only a few pixels in the image of the galaxy taken by its charge-coupled device camera, but this was enough to indicate that a powerful supernova was flaring up. Those few hundred photons were just enough

The photomultiplier

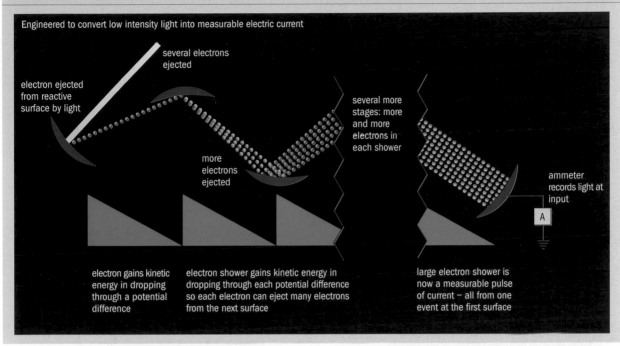

Engineered to convert low intensity light into measurable electric current

several electrons ejected

electron ejected from reactive surface by light

more electrons ejected

several more stages: more and more electrons in each shower

ammeter records light at input

A

electron gains kinetic energy in dropping through a potential difference

electron shower gains kinetic energy in dropping through each potential difference so each electron can eject many electrons from the next surface

large electron shower is now a measurable pulse of current – all from one event at the first surface

Clicks randomly spaced in time

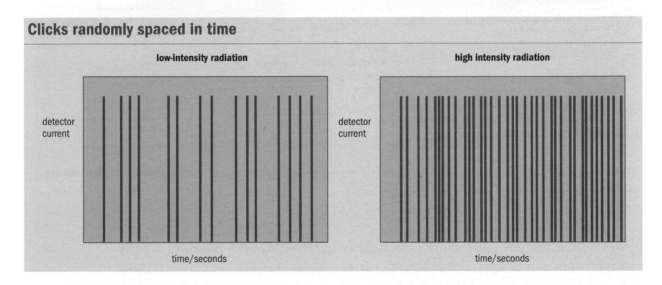

low-intensity radiation

detector current

time/seconds

high intensity radiation

detector current

time/seconds

to be distinguished from a random fluctuation in the other photons coming in from the galaxy.

Your eyes are excellent detectors of photons: when dark-adapted, the rods in the retina can just about detect single photons. The lumpiness of light can be shown experimentally using a photomultiplier that amplifies the arrival of a single photon to become a measurable pulse of electrons, so that incoming photons can be counted one by one.

Gamma-ray photons arrive in more energetic quanta than those of visible light. Using a simple

detector, gamma-rays from a radioactive source can be heard arriving as distinct clicks. As the source is brought closer to the detector, the clicks merge into a continuous noise.

Listening to gamma-rays arrive you hear random clicks. Sometimes there are several clicks close together, sometimes there are longer gaps between clicks. The arrival of one quantum is completely independent of the arrival of the one before or after – there is no predetermined timetable for them. This randomness is found everywhere in quantum behaviour, for example

Key summary: evidence for the graininess of light

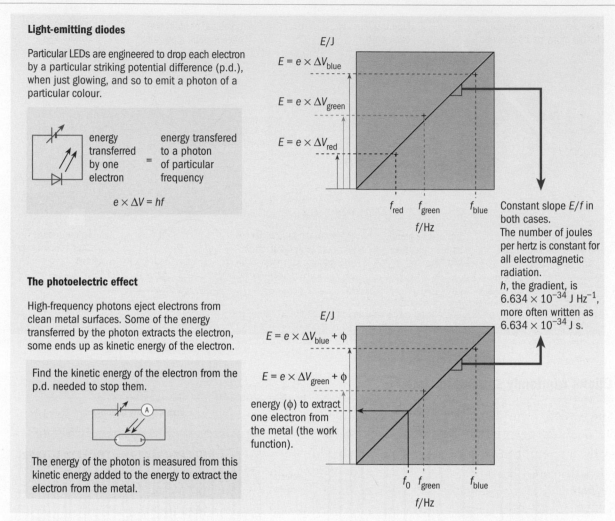

Light-emitting diodes

Particular LEDs are engineered to drop each electron by a particular striking potential difference (p.d.), when just glowing, and so to emit a photon of a particular colour.

energy transferred by one electron = energy transfered to a photon of particular frequency

$$e \times \Delta V = hf$$

$E = e \times \Delta V_{blue}$

$E = e \times \Delta V_{green}$

$E = e \times \Delta V_{red}$

E/J

f_{red} f_{green} f_{blue}

f/Hz

Constant slope E/f in both cases.
The number of joules per hertz is constant for all electromagnetic radiation.
h, the gradient, is 6.634×10^{-34} J Hz^{-1}, more often written as 6.634×10^{-34} J s.

The photoelectric effect

High-frequency photons eject electrons from clean metal surfaces. Some of the energy transferred by the photon extracts the electron, some ends up as kinetic energy of the electron.

Find the kinetic energy of the electron from the p.d. needed to stop them.

The energy of the photon is measured from this kinetic energy added to the energy to extract the electron from the metal.

E/J

$E = e \times \Delta V_{blue} + \phi$

$E = e \times \Delta V_{green} + \phi$

energy (ϕ) to extract one electron from the metal (the work function).

f_0 f_{green} f_{blue}

f/Hz

Both light-emitting diodes and the photoelectric effect give direct evidence for photons carrying energy in lumps $E = hf$

---

in the emission and detection of alpha and beta radiation from radioactive sources. Beneath events that seem to happen smoothly lie events happening one by one at random with a certain probability. Apparent smoothness is down to averaging. It is a bit like going out in the rain. You can't tell when or where the next raindrop will fall but, if you stay outside, you will get wet. Strong sources close up – like heavy rain – give a rapid average rate of arrival, while for weak ones far away – like a light shower – photons come less often. This randomness extends to the whole of quantum physics. Atoms emit photons at random, nuclei decay at random, but always underneath is a fixed probability that can be calculated.

Frequency and quantum energy

It was Max Planck (1858–1947) who first wrote down the connection between a quantum of energy of electromagnetic radiation and its frequency. The relationship is very simple. Electromagnetic radiation of frequency f is emitted and absorbed in quanta of energy E where

$$E = hf$$

in which h is the Planck constant, 6.6×10^{-34} J s.

The Planck constant h relates the quantum energy to the frequency of the radiation. Higher-frequency radiation comes in quanta of larger energy. The quantum energy of X-rays or gamma-

Key summary: how to do quantum calculations

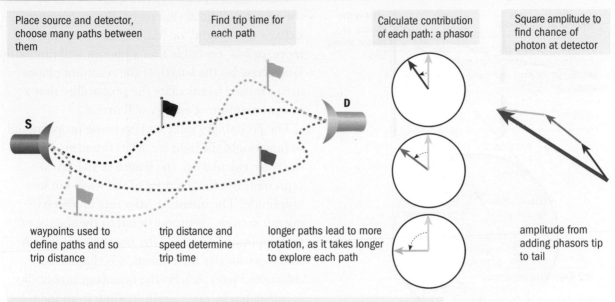

| Place source and detector, choose many paths between them | Find trip time for each path | Calculate contribution of each path: a phasor | Square amplitude to find chance of photon at detector |

waypoints used to define paths and so trip distance

trip distance and speed determine trip time

longer paths lead to more rotation, as it takes longer to explore each path

amplitude from adding phasors tip to tail

Repeating for more paths produces more accurate calculations. With limited time: choose wisely.

Add the phasors for all possible paths to find the resultant phasor for the whole process

rays is big enough to knock an electron out of an atom, which is why they can damage living tissue. The quantum energy of radio waves is tiny, which is why at most they stir electrons in metal aerial rods into gentle motion.

When Planck stumbled on this idea he was trying to "fix" the awkward fact that the wave theory of light gave the absurd result that even slightly warm objects should emit X-rays. But he was never really happy with it. A century later, however, quantum ideas are essential to a physicist's everyday work. Atoms and molecules emit or absorb photons of characteristic energy and frequency. Their sharp spectra are used in: monitoring car exhausts; finding the rate of rotation of stars; analysing the pigments in a mediaeval painting; identifying substances found at the scene of a crime; mapping a patient's internal organs in a body scanner.

Steal the wave calculations; forget the waves

Photons do something else besides arriving and departing in discrete quanta of energy. When there are alternative paths to the same point, interference effects appear. Chapter 6 is full of examples of superposition – thin films, Young's slits, gratings, diffraction. Superposition must be

fundamental to quantum behaviour.

Here we will take one of the simplest possible illustrations of quantum superposition: Young's double-slit experiment. There are just two paths – going by one slit and going by the other. The rule of quantum behaviour says: "Do both!". The photon explores both paths.

This sounds peculiar if the story of photon energy arriving in lumps has led you to imagine a photon as a little parcel travelling like a particle. Such a parcel can't be in two places at once. So a photon, which explores both paths at once, can't be like a little localised parcel. Lumpiness in energy doesn't mean lumpiness in space. A photon can't simply be a particle.

Waves have no problem in being spread out in space; they send wavelets through both slits, which superpose at the screen. But waves don't deliver energy in well-defined lumps so a photon can't simply be a wave. Just because photons go everywhere possible doesn't mean that their lumpy energy is spread thinner and thinner.

Quantum behaviour isn't particle behaviour and it isn't wave behaviour. Quantum objects (like photons) explore all possibilities, and yet still arrive in lumps.

Key summary: double-slit photon experiment

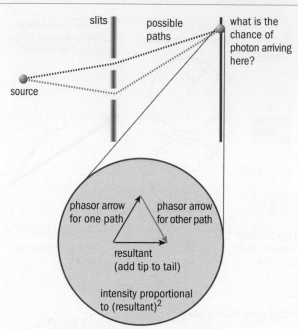

Add the phasors for the two paths to get the resultant for both paths

Wave calculations give the right answers for where there are bright and dark fringes on a screen when Young's double-slit experiment is performed using photons. So try a bold idea: steal the method of calculation from the wave story, but don't let that make you think that the photons are waves. Just borrow what you need, nothing more.

What then is actually used in Young's slits calculation? Surprisingly little. There are two paths, one from each slit. There is a difference in the lengths of the paths, which gives a difference in phase for light coming along each path and decides whether the screen will be light or dark. A phasor arrow can keep track of the phase at the end of each path.

How fast should the phasor arrow turn? What frequency f should it be given? Easy: adapt the relationship to photon energy E found by Planck:

$$f = \frac{E}{h}$$

That's all. No waves as such, only path difference and phase difference. Just two spinning phasor arrows that end up pointing in the same or different directions, and must be combined tip to

tail into a single resultant.

The resultant phasor decides how bright or dark the screen will be at that place. But in quantum behaviour, "bright" or "dark" just mean that it is more, or less, probable that a photon will randomly land there. So the length of the resultant phasor arrow is used to calculate the probability that a lumpy quantum of energy will arrive.

The probability must end up being proportional to how bright the light is – to its intensity. In the wave calculation, the length of the arrow represents not the intensity of the wave but its amplitude. The intensity – the rate at which energy arrives – is proportional to the square of the amplitude, which is why big waves at sea are so dangerous (for the reasons, see chapter 10, *Advancing Physics A2*). So the quantum probability is calculated by once again stealing from the wave calculation – get the probability by squaring the length of the arrow. So if the resultant phasor at one point is three times as long as the phasor at another, the probability of photons arriving is nine times as great at the first point than at the second.

Quantum behaviour for photons

Quantum behaviour is unlike anything else. For photons its rules are as follows.

- A photon is emitted by a source and is detected at a certain place and time.
- Imagine the photon taking every possible path from the source to arrive at that place and time. The longer the path, the earlier the time at which the photon will have to be emitted.
- A path includes the whole process of emission, travel and detection of a photon. For each such path, a combined phasor arrow can be calculated. This phasor arrow can be thought of as rotating at frequency $f = \frac{E}{h}$ for a time equal to the trip time of the photon. Thus the angle at which the phasor arrow ends up can be determined for each path.
- Add up the phasor arrows for all possible paths, tip to tail, to get their resultant phasor.
- From the square of the resultant phasor, calculate the probability of a photon arriving.

The phasor sum describes how the different possibilities combine together to give a probability. That's all there is to it – not wave behaviour, not particle behaviour, but quantum behaviour.

Quick check

1. Single photons of frequencies $f = 3.0\,\text{GHz}$, $f = 6.0 \times 10^{14}\,\text{Hz}$ and $f = 1.2 \times 10^{18}\,\text{Hz}$ are absorbed. Show that the energies transferred are $1.8 \times 10^{-24}\,\text{J}$, $3.6 \times 10^{-19}\,\text{J}$ and $7.2 \times 10^{-16}\,\text{J}$ respectively.

2. Assuming that low-energy light bulbs are 20% efficient and that the average wavelength of the light they emit is $5 \times 10^{-7}\,\text{m}$, show that the number of photons produced each second from a typical 20 W bulb is about $10^{19}\,\text{s}^{-1}$.

3. Show that the rate of rotation for the phasor associated with a photon of energy $3.3 \times 10^{-19}\,\text{J}$ is $5.0 \times 10^{14}\,\text{Hz}$.

4. A photon of frequency $6.0 \times 10^{14}\,\text{Hz}$ explores a path 6 m long. Show that its travel time is $2.0 \times 10^{-8}\,\text{s}$ and that its phasor rotates 1.2×10^{7} times.

5. Show that the frequency of a photon whose phasor rotates once exploring a path 200 mm long is 1 GHz and that its wavelength is 0.2 m. Would it be possible to set up laboratory-bench-sized superposition demonstrations using such photons?

6. In a Young's double-slit experiment, the paths from the slits to a place on the screen differ by 900 nm. If its frequency corresponds to a wavelength of 600 nm, show that the photon phasor makes 1.5 extra turns on the longer path and that the screen will be dark at that place.

Links to the *Advancing Physics* CD-ROM

Practise with these questions:

10S Short answer *Rotations for exploring paths*

20E Estimate *Photons streaming from a lamp*

30S Short answer *Path lengths and arrow rotations*

Try out these activities:

10E Experiment *Relating energy to frequency*

20D Demonstration *Listening to photons arriving*

70S Software-based *Many paths explored at once*

Look up these key terms in the A–Z:
Interference of photons; phase and phasors; photons; quantum behaviour; probability

Go further for interest by looking at:

10W Reading: well actually/but also *Interference of photons: the "many paths" story*

Revise using the revision checklist and:

20S Computer screen *Images with increasing exposure*

300 OHT *A path contributes an arrow*

400 OHT *Finding probabilities*

7.2 Many paths at work

"Explore all paths" sounds like a recipe for anarchy, not order. But it does have a definite rule – add up the phasor arrows for all the paths. And that gives new insights into many of the simple things that you already know about light.

Looking-glass world

Look in a mirror. Photons come off the mirror surface at the same angle as the angle at which they reach it. No quantum anarchy here, it seems. Not quite so. The "explore all paths" rule actually explains why that special angled path is favoured.

Think of a source of light and a detector above a mirror. Take a photon from source to detector by any path that hits the mirror. Yes, any path, including the ones hitting the mirror near the ends, which you know "don't happen". On each path, record the trip time in rotations of a phasor,

turning at frequency $f = \dfrac{E}{h}$, with speed c for the photon. A handy model for this is a trundle wheel that you roll along each path.

You will notice that there is something special about the paths near the one where the angle of reflection is equal to the angle of incidence. Phasors for these paths differ little in phase at the detector, and so they more or less line up. But phasors going by other, "wrong" paths differ a lot in phase from nearby ones, so that when placed tip to tail they tend to curl up. The lined-up phasors give a big resultant; the curled-up phasors contribute little to the resultant.

This gives you a new story about light, which turns what you have learned before on its head. You probably learned the law of reflection; quantum behaviour provides an explanation of that law. The rule of quantum behaviour sends photons everywhere, not just along one special

Key summary: reflection – explorations over a surface

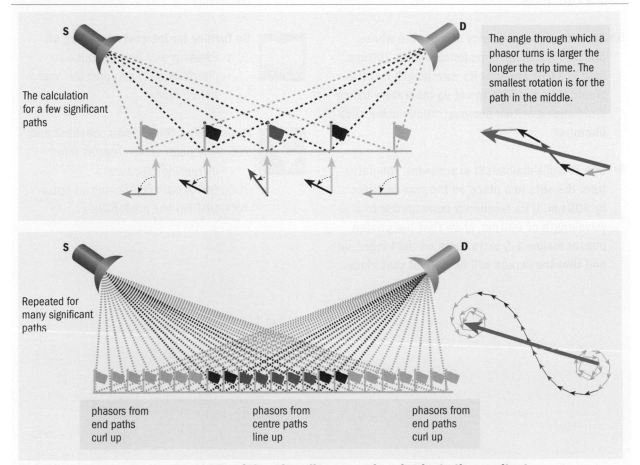

The calculation for a few significant paths

The angle through which a phasor turns is larger the longer the trip time. The smallest rotation is for the path in the middle.

Repeated for many significant paths

phasors from end paths curl up

phasors from centre paths line up

phasors from end paths curl up

Only phasors for paths via the middle of the mirror line up, and so dominate the resultant

Key summary: trip times, phase differences

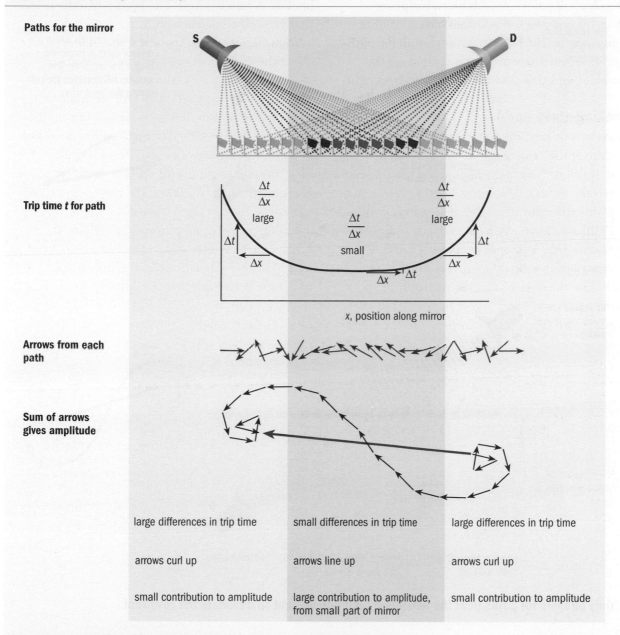

Paths for the mirror

Trip time *t* for path

Arrows from each path

Sum of arrows gives amplitude

large differences in trip time	small differences in trip time	large differences in trip time
arrows curl up	arrows line up	arrows curl up
small contribution to amplitude	large contribution to amplitude, from small part of mirror	small contribution to amplitude

Arrows line up where the trip time hardly changes as the path varies

path. However, the special equal-angled path obeying the law of reflection is in fact picked out by the quantum rule for combining phasors from every possible path. This special path, and ones very close to it, are the only ones for which the phasors line up rather than curl up. Most of the probability of photons arriving at the detector comes from paths fitting closely to the law of reflection. The rules of quantum behaviour predict the law of reflection.

Flat valley bottoms

The paths near to where the trip times are equal and the phasors line up are also the shortest paths – the ones with the shortest trip time. A graph of trip time has a minimum – it forms a valley. And any valley is flat at the bottom. If the trip time is a minimum, changing the path a little hardly changes the time. Result: the phasors from these paths line up. Quantum behaviour explains both reflection and Fermat's least-time principle (p138).

Key summary: refraction – explorations through a surface

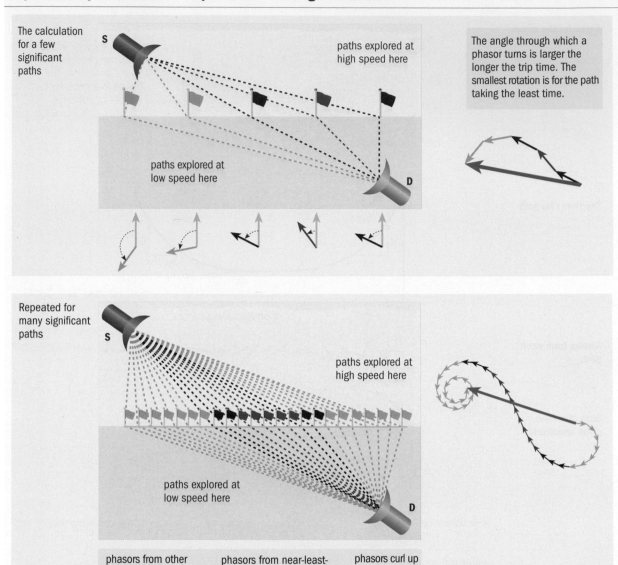

The calculation for a few significant paths

paths explored at high speed here

paths explored at low speed here

The angle through which a phasor turns is larger the longer the trip time. The smallest rotation is for the path taking the least time.

Repeated for many significant paths

paths explored at high speed here

paths explored at low speed here

phasors from other paths curl up

phasors from near-least-time paths line up

phasors curl up

Only phasors for paths near the least-time path line up, and so dominate the resultant

Which is the quickest path to get to the swimmer?

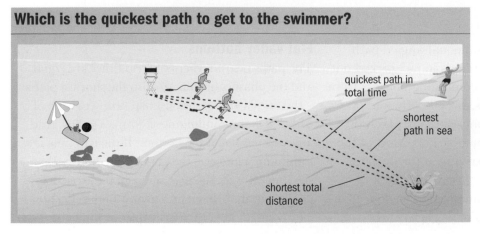

quickest path in total time

shortest path in sea

shortest total distance

A lifeguard is trying to get to a bather in trouble in the shortest possible time. The lifeguard can run quickly on the beach but can only swim slowly in the sea. The best path is not the direct path as this travels farther in the sea than is necessary. The best path is bent, going farther on the beach and taking a shorter line in the sea.

Key summary: propagation – explorations over space

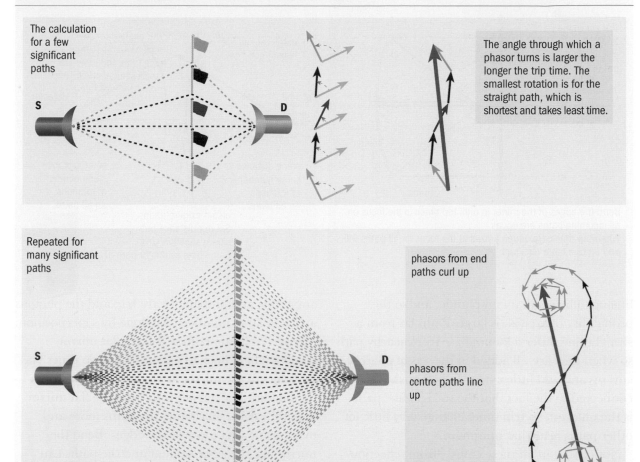

The calculation for a few significant paths

S   D

The angle through which a phasor turns is larger the longer the trip time. The smallest rotation is for the straight path, which is shortest and takes least time.

Repeated for many significant paths

S   D

phasors from end paths curl up

phasors from centre paths line up

phasors from end paths curl up

Only phasors for paths near the least-time straight-line path line up, and so dominate the resultant

Going faster, going slower

Chapter 6 shows how Huygens explained refraction in terms of waves slowing down in water or glass. The photon story keeps the slowing down, but once again says to a photon, "try all paths". Because of the difference in speed, trip time is now important.

The photons do not sit down beforehand and work out what to do. As usual, they just try everything. But on the paths near the one of shortest time, the phase differences between paths are all small and so the phasors for these paths line up. Paths around those farther from the least-time path have big phase differences between them, and the sets of phasors from these curl up when added in. So once again, the anarchic-seeming but actually strict rules of quantum behaviour predict a definite result:

most of the probability for photons arriving comes from paths close to the observed refracted path.

So refraction is yet another consequence of the quantum behaviour of photons.

Seeing straight

It's a remarkable fact that reflection and refraction turn out, in quantum physics, to be examples of superposition, that is, of interference. It is even more remarkable to realise that so is the fact that you can't see round corners.

You may well have guessed the story by now: that light goes in straight lines because photons try every path, including paths that are not straight. You would be right. Paths near the straight-line path differ hardly at all in phase. Their phasors

Engineering a mirror with phasors

Bend the edges of the mirror in until trip times to the focus on all incoming paths are equal
When trip times are equal, arrows at the focus for all paths will add up to a large resultant

Engineering a lens with phasors

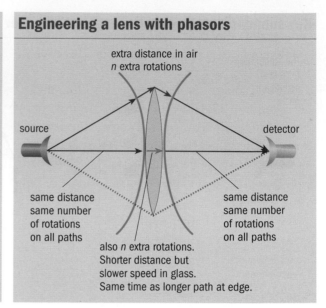

extra distance in air
n extra rotations

source

detector

same distance
same number
of rotations
on all paths

same distance
same number
of rotations
on all paths

also *n* extra rotations.
Shorter distance but
slower speed in glass.
Same time as longer path at edge.

line up. The resultant amplitude, and so the probability of arrival, is large. Paths far from a straight line differ a lot in phase from nearby paths so when they are all added in the sets of phasors curl up and add little extra. As before, this is connected to the fact that the straight-line path is the quickest, so trip times change very little for other paths near that minimum.

Notice that in all these cases – from reflection to travelling in a straight line – you always have to think about bundles of paths all close to a given direction. If the phasors for the paths in a bundle line up, that bundle makes a big contribution to the resultant. If they curl up, that bundle makes little or no contribution. Thus in the present case – straight line propagation – only the bundle of paths close to a straight line from source to detector is important in getting photons from source to detector with high probability. The bundle is very narrow: all paths in it must differ in length by less than a fraction of a wavelength.

Engineering with photon paths

A telescope designer or the optical engineer of a DVD player wants to get a lot of photons from a source to come together at the same place. The photons don't care: they go everywhere. All the designer has to do is to arrange it so that the trip times along as many different paths as possible are the same. Then the phase differences for those paths are small; their phasors line up; the

amplitude and probability are big; and the photons arrive in large numbers. All done by superposition.

Easy? Well, sometimes. Take a flat mirror reflecting light back to a point. The trip times via the edges of the mirror are longer than those via the middle. The answer is to bend the flat mirror in at the edges, shortening the paths more and more as they get towards the edges. Bend the mirror just the right amount and the paths can all be the same length. That's why telescopes use curved mirrors.

How about spectacle lenses or a magnifying glass? The optical engineer must stop light travelling in straight lines and get it to bend. The paths far from the straight one take too much time. They can't be made shorter, but instead the straight-through paths can be made slower to get the same effect. How? By putting more glass in the way of straight paths than of up-and-down paths. The solution you know already – a convex lens. All paths through the lens take the same time to go from object to image. The longer ones travel a shorter distance through slow-speed glass so the light arrives in phase from them all.

Getting more from less

The motto for superposition is, "more can mean less". Equally true is, "less can mean more". Here we explain how getting rid of possible paths can increase the number of photons going in a certain direction. Joseph Fraunhofer did it when he made the first diffraction grating (p142). It sent light of a particular wavelength off in a direction that it would not have gone in before, a new direction in which all the path phasors line up.

Think about the story for light travelling in straight lines. The off-centre paths curl up and contribute little. That's because they alternately point one way and then the opposite way. So blocking off alternate paths leaves only phasors pointing mainly the same way, producing a bright beam of light.

If you look at chapter 6, p143, you'll see that this is another way of telling the story about how a grating works.

Really going everywhere

Look at a small bright lamp through the crack between two of your fingers, and gently squeeze the fingers together. Just before they close, the light blurs out sideways. Angles where there was no chance to see a photon suddenly get some chance. This is called diffraction (chapter 6, p146). It happens because photons really do arrive with a probability that takes account of everywhere they might have gone.

Imagine a slit so narrow that all paths from any part of the slit to anywhere else are pretty much the same length, counted in phasor turns. For that, the slit must be narrower than the distance for one phasor turn – one wavelength. Then photon paths at all angles after the slit will have much the same probability for a photon to arrive along one path as along any other. The light will spread out into a uniform blur.

Open up the slit and more paths become possible. Some directions have phasor resultants that are small or zero. The beam behind the slit gets narrower. Open the slit completely and you are back to light travelling in straight lines, because all the other paths add up to zero amplitude.

The result is that the way to make a narrow beam of photons is the opposite of what you might think. Narrow the beam down and it spreads. Open it up and the photons go straight. Astronomers and satellite engineers know it well: to get good directionality, make telescopes and aerials as wide as possible. That's one of the things that makes astronomy expensive.

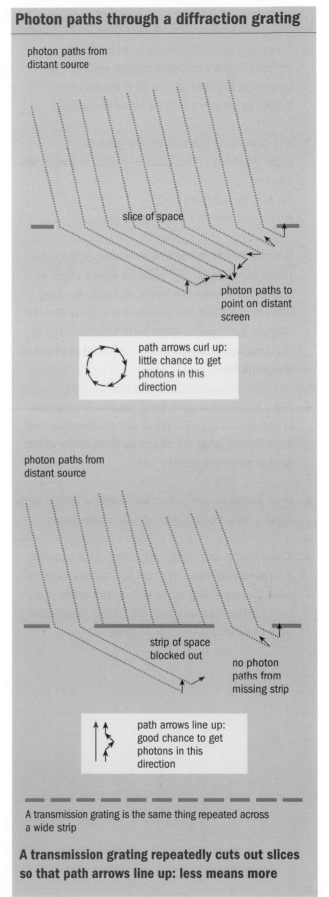

Photon paths through a diffraction grating

photon paths from distant source

slice of space

photon paths to point on distant screen

path arrows curl up: little chance to get photons in this direction

photon paths from distant source

strip of space blocked out

no photon paths from missing strip

path arrows line up: good chance to get photons in this direction

A transmission grating is the same thing repeated across a wide strip

A transmission grating repeatedly cuts out slices so that path arrows line up: less means more

Quick check

1. Three photon paths in air from source to detector have the same length. Here is the phasor at the end of one of the paths: →. Draw the phasors for the other two paths.

2. The phasor at the end of one path is →. Draw phasors at the ends of paths on which the phasor:
   (a) makes one extra turn;
   (b) makes two and a half extra turns.

3. Given a fixed source and detector, paths via a plane mirror near the path for which angle of reflection is equal to angle of incidence vary little in length if the angles are slightly altered. Why do such paths contribute substantially to the resultant amplitude for arrival of photons at the detector?

4. The straight-line path from source to detector is the shortest path. What effect does this fact have on the phasors at the ends of paths close to this shortest path?

5. Give a reason why a parabolic mirror brings light from a very distant source to a sharp focus.

6. Show that a diffraction grating with a spacing d between its lines of 1.2×10^{-6} m will have its path arrows lining up to give a first order (i.e., $n = 1$) diffracted beam at 30° when illuminated with light of wavelength 600 nm.

Link to the *Advancing Physics* CD-ROM

Practise with these questions:

40S   Short answer *Three paths on a mirror*

60X   Explanation–exposition *Spacing a grating*

70X   Explanation–exposition *CD: mirror and grating?*

90C   Comprehension *Mirrors for precision engineering*

140S Short answer *A quantum view of diffraction*

Try out these activities:

110S Software-based *A few mirror paths*

120S Software-based *Many photons make a beam*

170S Software-based *Engineering a focusing mirror*

Look up these key terms in the A–Z:
Diffraction; gratings and spectra; phase and phasors; quantum behaviour; reflection; refraction

Go further for interest by looking at:
30W Reading: well actually/but also *The principle of least action*

Revise using the revision checklist and:
900   OHT *Mirror: contributions from different paths*

1000 OHT *Calculated trip times for a mirror*

7.3 Electrons do it too

We said at the beginning of this chapter that quantum behaviour is important because it is common to all the fundamental particles you may read about, including neutrinos and quarks. Here we will extend the ideas to electrons.

An electron is not exactly like a photon. Electrons have mass, which means that they cannot travel at the speed of a photon (the speed of light) but they can travel at different speeds slower than light. But, just like a photon, an electron offered more than one possible path explores them all. Just like a photon, a rotating arrow or phasor tracks the phase change along every path, and these arrows superpose in just the same way.

It took some ingenuity to repeat Young's classic double-slit experiment to show that electrons can interfere just like photons. To see electron quantum behaviour it is easiest to use atom-sized spacings of slits. Atoms arranged in regular rows can be used as a natural finely spaced grating.

It's easy to make the electrons go faster or slower by changing the accelerating potential difference. If you do, the diffraction pattern changes. The rate at which the phasor arrow for an electron path turns depends on the energy of the electrons.

Rotating electron arrows

The quantum behaviour of electrons is similar to that of photons (p156). But we now have to include the mass m and speed v of the electrons in the rules. For electrons moving at much less than the speed of light, the description of quantum behaviour becomes:

- an electron of mass m, speed v is emitted by a source and is detected at a certain place and time;
- imagine the electron taking every possible path from the source to arrive at that place and time. The longer the path, the earlier the time at which the electron will have to be emitted. How much earlier depends on the speed of the electron;
- a path includes the whole process of emission, travel and detection of an electron. For each such path, a combined phasor arrow can be calculated. The phasor arrow can be thought of as making one turn for each distance $\frac{h}{mv}$ along the path, thus the angle at which the phasor arrow ends up can be determined for each path;

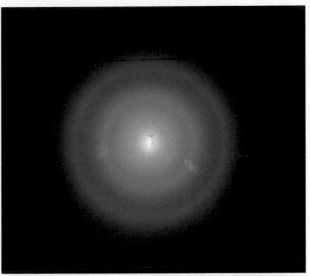

This image shows an electron diffraction pattern generated when electrons fired at a thin sheet of graphite are diffracted by the regular rows of carbon atoms.

Electrons really do interfere like photons. These are photographs of electron diffraction at single (left) and double (right) slits.

- then, exactly as for photons, add up the phasor arrows for all possible paths, tip to tail, to get their resultant phasor, and from the square of that resultant phasor, calculate the probability of detection of an electron.

The quantity $\frac{h}{mv}$ for the path distance for one turn of the arrow was introduced by French physicist Louis de Broglie when quantum ideas were first being developed in the 1920s. He thought of it as like a wavelength λ, with

$$\lambda = \frac{h}{\text{momentum}}$$

where the quantity mv is named momentum.

This relationship can be checked experimentally by varying the speed of the electrons in an electron diffraction tube and seeing how the diffraction pattern changes. To vary the electron speed you just increase or decrease the accelerating potential difference in the tube.

To see that the electron rule is actually just like the photon rule, you have to know that photons also have momentum (chapter 16, *Advancing Physics A2*). Their momentum is just $\frac{E}{c}$, where c is the

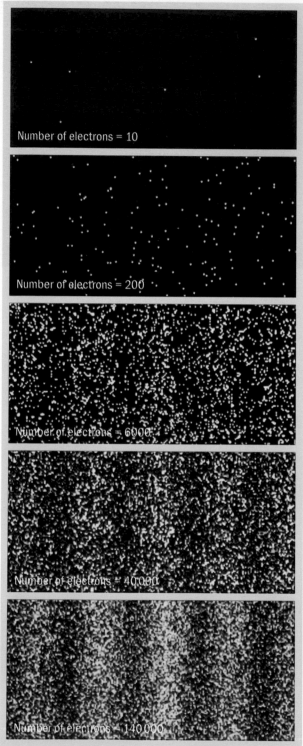

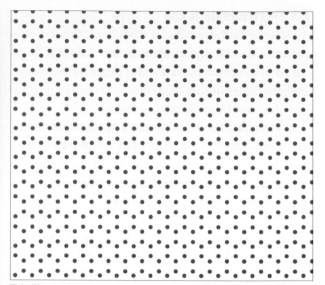

This illustration shows the arrangement of atoms in graphite. Pick up the book and look edgeways at the diagram. You will see densely packed rows. If you turn the page a little you can see other rows.

These remarkable images were obtained by Akiro Tonomura. They show an electron interference pattern being built up as electrons arrive one by one at a detector. There are two alternative paths for the electrons, as in a double-slit experiment. At any moment, only one electron is passing through the apparatus. A spot of light is generated where it arrives at the detector. The series of images show the pattern of arrival of electrons as the number of electrons arriving increases. The fact that the electrons arrive randomly, with a probability greatest where the fringe pattern is most intense, is clearly demonstrated.

photon speed. Because in a second the photon travels distance c and the photon arrow makes $\frac{E}{h}$ turns, the distance for one turn of the arrow is given by:

$$\text{distance for one turn} = \frac{\text{distance per second}}{\text{turns per second}}$$

$$= \frac{c}{E/h} = \frac{hc}{E}$$

Thus:

$$\text{distance for one turn} = \lambda = \frac{hc}{E} = \frac{h}{(E/c)} = \frac{h}{\text{momentum}}$$

You see that this is exactly the same expression as for electrons.

Recently, Akiro Tonomura of Hitachi has been able to detect electrons arriving one by one, building up a two-slit interference pattern, with many electrons arriving where the probability is high and few arriving where it is low. First he built a highly monoenergetic source of electrons. The electrons passed through a gap between two metal plates, passing either side of a thin negatively charged wire between the plates, and were then brought together again at a detector. The place where each electron arrived was recorded. Fringes similar to those in a double-slit experiment were seen in the pattern of arrival of the electrons.

Quick check

Useful data: Charge on electron $e = -1.6 \times 10^{-19}\,C$, mass of electron $m_e = 0.9 \times 10^{-30}\,kg$, Planck constant $h = 6.6 \times 10^{-34}\,Js$.

1. Show that the kinetic energy of electrons accelerated by a p.d. of 100 V is $1.6 \times 10^{-17}\,J$.

2. Show that the speed of the electrons in question 1 is about $6 \times 10^6\,m\,s^{-1}$.

3. Show that the momentum mv of the electrons in question 1 is about $5 \times 10^{-24}\,kg\,m\,s^{-1}$.

4. Show that the de Broglie wavelength for these electrons is about the same size as a small atom ($10^{-10}\,m$).

5. What problem would there be trying to do a double-slit superposition demonstration with such electrons?

6. Show that the speed and de Broglie wavelength of an electron with an energy of 2 eV (similar to a photon of visible light) would be about $10^6\,m\,s^{-1}$ and 1 nm.

Links to the *Advancing Physics* CD-ROM

Practise with these questions:
145S Short answer *Electron volts – a new unit*
160S Short answer *Electrons through gratings*

Try out this activity:
300S Software-based *Electrons interfering one by one*

Look up these key terms in the A–Z:
Electron diffraction; quantum behaviour

Go further for interest by looking at:
20W Reading: well actually/but also *Interference of electrons: the "many paths" story*

Revise using the revision checklist

7.4 What does it all mean?

A peculiar thing about quantum behaviour is that although everyone agrees how to work things out, and everything agrees with experimental results, there is more than one picture of what lies behind the calculations and no agreement about which is the "right" picture. That's just how it is. So we have chosen the story that seems to be the simplest, a story first worked out by physicist Richard Feynman. This was part of a development of quantum ideas around 1945–50 that won him, his fellow American Julian Schwinger and Japanese physicist Sin-Itiro Tomonaga the Nobel prize. Tomonaga got there first, even in war-torn Japan.

You mustn't suppose that the picture we have painted is exactly "how things are". The story of photons following all possible paths is an easy way of remembering how to do the calculations, but don't let it carry you away. For example, don't imagine photons trying paths one at a time, one after the other. If anything, they must be imagined trying them all at once.

Feynman's "try all paths" idea is not altogether new. Huygens (chapter 6) long ago sowed the seeds. He thought of his wavelets as spreading out everywhere they could, and building up the new position of a wave by arriving there all in phase. The wavelets go everywhere, but in most places add up to nothing at all.

Shocking or not?

People originally found quantum behaviour peculiar, even shocking. Many books have been written about its weirdness. You will have to decide for yourself, once you have got used to it.

It does seem peculiar that quantum behaviour steals the mathematics of waves, without supposing that waves lie behind it. But let us try to turn this question round: why is a wave theory of light possible if photons are quantum objects?

Here's an answer, see what you think of it. Quantum behaviour is governed by combining phasors (amplitude and phase in one spinning arrow) to calculate probabilities. Suppose that this is all there is to say about how photons work. What happens when there are countless numbers of photons in a beam? The probability just

Richard Feynman (right) was renowned for his lightning-speed calculating ability and for his "no nonsense" approach to problems in physics and in life.

Neils Bohr on quantum theory

"Anyone who is not shocked by quantum theory has not fully understood it."

determines the brightness, which varies smoothly from place to place. And the brightness varies just as if waves were superposing to give the result, because of the quantum adding-up of phasors.

This turns the usual argument on its head. The usual argument says that photon behaviour is calculated by wave-like methods because photons are a bit like waves. Our argument says that light is a bit like waves because photons show quantum behaviour. Waves are what photons appear to be when the photons are numerous.

This argument may feel uncomfortable. Something almost obvious is being explained in terms of something strange. Shouldn't it be the other way round? But if so, how will a familiar thing ever get explained?

Waves or particles?

The modern story we have told avoids the puzzle: "Are these things waves or particles?" Many a book about quantum physics scratches this ancient sore spot. Result: it just itches more. Our answer is that electrons and photons are neither – they are just themselves, quantum objects.

This doesn't mean that the puzzles go away. Photons somehow manage to go everywhere and yet always be detected somewhere. Arguments about the consequences of this "non-local" yet "localised" behaviour continue to this day. They are leading to new ideas about quantum computing which, if they can be made to work, would be immensely powerful.

Summary check-up

Quantum behaviour ✓

- Quantum behaviour is unique; it is neither wave nor particle behaviour. Its fundamental rule is "try all paths".

Photons ✓

- Photons show quantum behaviour
- The energy delivered by a photon is $E = hf$
- Quantum behaviour combines phasors from all possible paths
- The probability of an event is found from the square of the resultant phasor amplitude
- The rate of rotation of the phasor for photons is simply the frequency of the light involved

Paths taken by photons ✓

- Photons will seem to go along special paths if near those paths the phasor arrows line up, while elsewhere they curl up
- Phasor arrows for photons line up for paths near the path that takes the least time
- These principles account for reflection, refraction and straight-line propagation, help in the design of curved mirrors and lenses, and account for diffraction from a slit
- Gratings get certain definite paths to have phasors that line up by removing paths that when added in make the phasors curl up

Electrons ✓

- Electrons also show quantum behaviour
- The wavelength associated with electrons is $\lambda = \frac{h}{mv}$ (at velocities much less than the speed of light)
- The phasor for a free electron travelling at speed v turns once for every distance $\frac{h}{mv}$ along a path
- Electrons produce superposition effects when passed through a grating that has spacing on an atomic scale

Questions

1. **Paths for photons travelling from a source to a detector via a mirror are shown. Three paths go via A near one edge of the mirror. Three paths go via B where the angle of incidence is equal to the angle of reflection. The graph shows how the trip time varies with the position of the point where the path meets the mirror.**

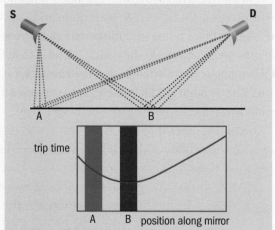

(a) Are the trip times for the paths near A similar?
(b) Are the trip times for the paths near B similar?
(c) Sketch the three phasors at the detector for the paths going near A.
(d) Sketch the three phasors at the detector for the paths going near B.

2. **Fill in the spaces in the tables below. (Planck's constant $h = 6.6 \times 10^{-34}$ J s; speed of light $c = 3.0 \times 10^8$ m s^{-1}; charge on electron $e = 1.6 \times 10^{-19}$ C; mass of electron $= 9.1 \times 10^{-31}$ kg; energy in joules = energy in electron volts $\times$ charge on electron.)**

Photons

Frequency of phasor rotation	Energy	Wavelength
-	-	300 nm
300 GHz	-	-
-	1.6×10^{-19} J	-

Electrons

Momentum mv	Speed v	Wavelength
-	-	10 nm
1.0×10^{-24} kg m s^{-1}	-	-
-	2.0×10^6 m s^{-1}	-

3. **Photons go from a source to a detector, as shown.**

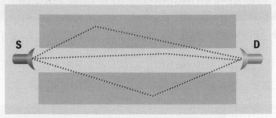

(a) Why do paths going through the shaded regions contribute little to the resultant phasor at the detector for all paths between source and detector?
(b) What has this to do with the rectilinear propagation of light?

4. **Light goes through a single narrow slit. Many photons arrive at a place at an angle to the direction of the light falling on the slit.**

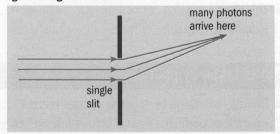

(a) Are the trip times of the three paths from the slit equal?
(b) If many photons arrive at the point shown, what can you say about the numbers of phasor rotations for the three paths?

5. **A converging lens is focusing light, as shown.**

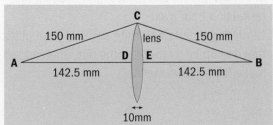

(a) Express the speed of light in millimetres per second (mm s^{-1}).
(b) How long does the light take to follow the route ACB?
(c) How long does the light take to cover the route ADEB?
(d) How long must the light spend between D and E in the glass?
(e) What is the speed of light in the glass?
(f) What is the refractive index of the glass?

8 Mapping space and time

Here we introduce you to a new kind of mathematical quantity – vectors. Vectors are a powerful way of describing arrangements in space, movements, forces, fields, among other things. We will give examples of:

● vectors in map-making
● vectors describing movement
● how to take vectors apart into components
● how to add vectors together

These ideas are important in all parts of physics and engineering.

8.1 Journeys

Robert Reid is a physicist who has studied cosmic rays at the South Pole. He describes his journey to the Pole: "The flight took eight hours in a Hercules aircraft, sitting in a cramped hammock seat, surrounded by cargo, with no heating and no insulation from the sound of the engines. It was too dark to read and the only sanitation was a bucket." The time the journey took, not the distance he travelled, is what sticks in his mind. You have probably felt exactly the same on a long trip.

Time is often used to describe distance. Bats listen to the echoes of the high-frequency chirps that they make, to know where objects are in front of them. Submarines use the same idea. Electronic tape measures from DIY stores use ultrasound

Robert Reid throws his favourite omega-shaped boomerang round the South Pole (marked by the board). This boomerang has a circular orbit of diameter 40 m and returned accurately directly behind the thrower. Time of flight? Choose for yourself between 9 seconds and 1 day and 9 seconds!

How far can you travel in eight hours?

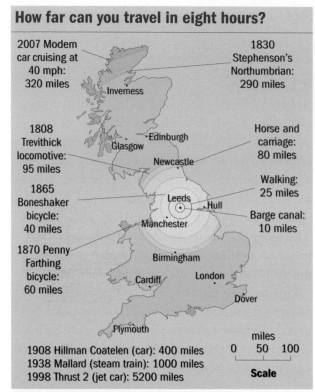

2007 Modern car cruising at 40 mph: 320 miles

1830 Stephenson's Northumbrian: 290 miles

1808 Trevithick locomotive: 95 miles

Horse and carriage: 80 miles

1865 Boneshaker bicycle: 40 miles

Walking: 25 miles

Barge canal: 10 miles

1870 Penny Farthing bicycle: 60 miles

Inverness, Edinburgh, Glasgow, Newcastle, Leeds, Hull, Manchester, Birmingham, Cardiff, London, Dover, Plymouth

miles
0   50   100
Scale

1908 Hillman Coatelen (car): 400 miles
1938 Mallard (steam train): 1000 miles
1998 Thrust 2 (jet car): 5200 miles

How far can you travel from Leeds in eight hours? The circles show the distances that could be travelled at the constant top speeds of these types of transport in eight hours. Why are these distances an overestimate?

A journey from Leeds to Manchester

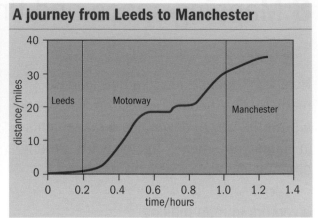

reflections to measure rooms; medical ultrasound scans locate babies in the womb (chapter 1).

To get from the journey time to the distance, you need to know the speed. If you can guess the speed of a Hercules aircraft, you can estimate the length of Robert Reid's trip. But the idea is deeper than you might think. The speed of light is special: it doesn't depend on how you are moving. So all journeys really can be described by time, the time light would take to make them. This is the starting point for Einstein's theory of relativity.

Transport

The past 100 years has seen a drastic change in how we travel. The distances we now travel every day would have been unthinkable a century ago. If we think about a day's travel (say eight hours on the road) as a sensible unit for measuring journey times, how has the distance that we can travel in this time changed from last century?

Today, we can travel huge distances in a day. Geneticist Steve Jones wrote, "…there is little doubt that the most important event in recent

human evolution was the invention of the bicycle". According to Jones it allowed mixing of the gene pool in ways that had not happened before its invention, when people typically married others in their own or very close communities. Transport has given people enormous personal freedom.

Speed

The route from Leeds to Manchester through the Pennine hills of the north of England has long been important for trade; it has carried a packhorse track, a canal, a railway and a motorway. The straight-line distance from Leeds to Manchester is about 35 miles; it takes the cross-country train an hour to travel this distance, a car about 75 minutes and a barge about 33 hours. In each case, the average speed for the journey is somewhat less than you might expect from an estimate of the speed of the type of transport. In a car, I can drive at 70 miles per hour along the motorway, so that the time to travel 35 miles is:

$$\text{speed} = \frac{\text{distance}}{\text{time}} \text{ or in symbols } v = \frac{s}{t}$$

so

$$\text{time} = \frac{\text{distance}}{\text{speed}}$$

or

$$\text{time} = \frac{35 \text{ miles}}{70 \text{ miles per hour}} = 0.5 \text{ hours}$$

Of course, the actual time is much longer than this because I cannot drive at this speed all the way; city-centre traffic soon sees to this, as does rush-hour traffic on the M62. A distance–time graph makes this clear.

Average and instantaneous speed

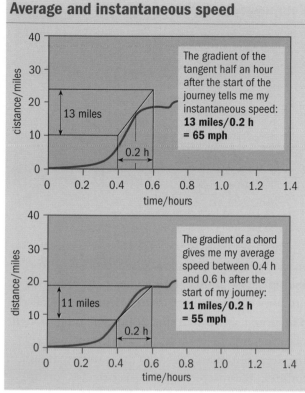

The gradient of the tangent half an hour after the start of the journey tells me my instantaneous speed:
13 miles/0.2 h = 65 mph

The gradient of a chord gives me my average speed between 0.4 h and 0.6 h after the start of my journey:
11 miles/0.2 h = 55 mph

Average and instantaneous speed can both be calculated from the gradient of a straight line.

In "A journey from Leeds to Manchester" opposite, the graph tells a story. Think about what is happening at each stage of the journey. Where was I stuck? Where did I have to slow down? In a sense, this graph is a map of the journey. A conventional road map lays out the route in two space dimensions. This map lays out the journey in a two-dimensional "space–time". It is a frozen story of the journey.

Instantaneous speed and average speed

At times, travelling from Leeds to Manchester in the car, I might drive at 70 mph, but at other times I sit in gridlock. What is my average speed? Easy! It is the total distance divided by the total time:

$$\text{average speed} = \frac{35 \text{ miles}}{1.25 \text{ hours}}$$

$$= 28 \text{ miles per hour}$$

Here is something odd: the speedometer in my car swings round to 70 mph as I move away from a traffic jam, but it does not need to wait until I have travelled for an hour to calculate this: so what is it

Distance–time graphs of a woman walking

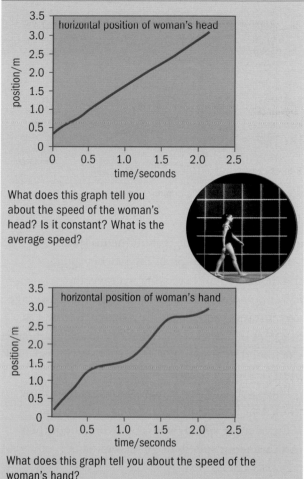

What does this graph tell you about the speed of the woman's head? Is it constant? What is the average speed?

What does this graph tell you about the speed of the woman's hand?

A motorway journey at constant speed

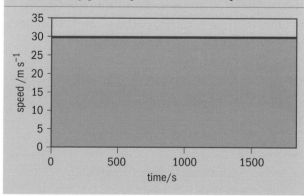

reading? It tells me the instantaneous speed of the car, the speed at that moment. In practice, I could estimate this by seeing how far the car travels in a very short space of time: this is what light gates do that you may use in practical work. The shorter the time interval, the more accurate is the

Distance travelled at varying speed

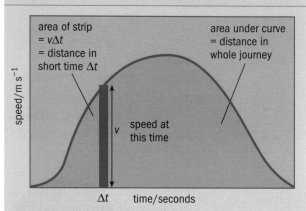

Journey with varying speed

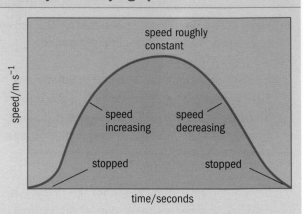

approximation of the instantaneous speed.

Let's look at a graph of my journey along the motorway and see what average and instantaneous speeds mean. The speeds – average and instantaneous – can be calculated from the gradient of a line. Notice that because a gradient is calculated by dividing a distance by a time its unit must be that of speed.

Have a look at the image of a woman walking p173. Her hand and her head must have the same average speed (why?) but what about their instantaneous speeds at given moments?

Speed–time graphs

Suppose on my journey from Leeds to Manchester along the M62 I am able to travel at a steady speed of 70 mph for 30 minutes. This time, let's convert to SI units: 70 mph is roughly 30 m s^{-1} and 30 minutes is 1800 s. How far will I have travelled?

$$\text{speed} = \frac{\text{distance}}{\text{time}}$$

So

$$\text{distance} = \text{speed} \times \text{time}$$

$$\text{distance} = 30 \text{ m s}^{-1} \times 1800 \text{ s}$$

$$\text{distance} = 54 \text{ km}$$

You've already thought about one way of picturing a journey: as a distance–time graph. Now look at a different picture: a speed–time graph.

In "A motorway journey at constant speed" (p173) the graph is very simple. Look at the

shaded part, a rectangle whose "area" is 30 m s^{-1} multiplied by 1800 s. This comes out as the quantity calculated earlier, the distance travelled. The graph shows that this distance can be thought of as the area below a graph. Notice the units of the area: not metres squared (m^2), the usual unit for area, but metres (m). This is because you are not multiplying a length by a length but rather a speed by a time. So watch out!

But your speed may not be constant. If you increase your speed, the graph of speed against time slopes upwards. If your speed decreases, the graph slopes downwards. The speed–time graph reaching zero means that you have stopped.

It's easy to see that the rectangular area under the graph for constant speed is the distance travelled, because distance = speed × time. But it isn't so easy to see that the area under any speed–time graph is the distance travelled, no matter how the speed changes.

One way to see it is to think of a short interval of time, say a few seconds. In that short time the speed can't change much, so there is a more-or-less definite speed v for that moment. The distance gone in the short time interval Δt is $v\Delta t$. This too is an area beneath the curve: it is the area of a thin strip, Δt wide, reaching up a height v to the curve.

Now imagine doing the same, strip after strip across the whole curve. The total area is the area of all the strips added together. The total distance travelled is the area under the curve.

This gives a useful general rule: the area below a speed–time graph between two times is the distance travelled between those times.

Quick check

1. Show that a Hercules aircraft moving at a speed of $250\,\mathrm{m\,s^{-1}}$ flies $7200\,\mathrm{km}$ in eight hours.

2. If I drive at 70 mph for 30 minutes followed by 30 mph for an hour, show that I travel a total of 65 miles and that my average speed is 43 mph. Why isn't my average speed simply the average of 70 and 30 (= 50 mph)?

3. Draw a distance–time graph for the journey in question 2. Draw a speed–time graph for this journey. Annotate the distance–time graph to show how the speed can be calculated. Annotate the speed–time graph to show how the distance can be calculated.

4. Show that I can walk about 60 km in eight hours if my average speed is $2\,\mathrm{m\,s^{-1}}$.

5. Show that the speed limit in urban areas of 30 mph is equivalent to $13.5\,\mathrm{m\,s^{-1}}$.

6. Look at the graph of my journey from Leeds to Manchester on p172. Show that my average speed for the journey was about 28 mph and that my instantaneous speed an hour and a half after starting was about 25 mph.

7. By thinking about the gradient of the graph, use the distance–time graph of my journey from Leeds to Manchester to sketch a speed–time graph. Annotate it with comments about what you consider to be the important features.

Links to the *Advancing Physics* CD-ROM

Practise with these questions:

10E Estimates *Making estimates about motion*

20S Short answer *Distance, time and speed calculations*

10D Data handling *A history of railway locomotives: speed, distance and time*

Try out these activities:

20S Software-based *Exploring distance–time graphs*

40S Software-based *Investigating motion graphs with the Multimedia Motion CD-ROM*

80S Software-based *Travel graphs with Modellus*

90S Software-based *Calculating speeds and distances from graphs*

Look up these key terms in the A–Z:
Distance–time graph; speed–time graph

Go further for interest by looking at:

10T Text to read *The metre takes the stage*

20T Text to read *Distance, time and speed in journeys*

Revise using the revision checklist

8.2 Maps and vectors

Only on a circular walking trip do you set out
to cover a big distance without going anywhere.
Usually on a journey you want to go a certain
distance in a particular direction. Distance in a
certain direction is called **displacement**, and it is
a vector. **Vector quantities** have both size, usually
called magnitude (distance in this case), and
direction. Vectors are very important in physics:
examples include velocity, acceleration and force,
which we shall consider later in this chapter and in
chapter 9. Quantities like speed, which have only a
magnitude, are called **scalars**. Volume is another
example of a scalar.

Suppose that you are on holiday in France. You
don't know the country, but you have a table of the
distances between towns ripped out of a road atlas.
What the table does not tell you is the direction
from one town to another. The table tells you
the scalar distance not the vector displacement.
It is possible to find the relative directions just
from the distances. Seeing how to do this helps to
understand how vector quantities combine.

Look at "Table of distances between French
towns". We'll consider the distances in red.

It is obvious from the figures, if you know what
to look for – even if you know nothing about the
geography of France – that Lyon lies very nearly
on a straight line between Calais and Marseille.

Not every pair of distances works out like this.
Look at the figures for Marseille, Bordeaux and
Lyon. These distances do not add up. The paths
must be angled.

To find a possible layout of these towns, lines
proportional to 426 and 498 km can be anchored
on Lyon and Marseille (see "Relative directions:
angled paths" opposite) and rotated until they meet.
(Note that the location of Bordeaux shown is not the
only possible one – circular arcs would also meet
above and to the right of Lyon and Marseille.)

What has this to do with vectors? A little
notation will help. Call the distance from
Bordeaux to Lyon a, and the distance from Lyon to
Calais b. Finally, call the distance from Bordeaux
to Calais c. If I write $a + b = c$ then you may tell
me I am wrong. But there is a sense in which
this is correct. If a now means the journey from
Bordeaux to Lyon (I have used a bold typeface,

Table of distances between French towns

	Bordeaux	Calais	Lyon	Marseille	Paris	Brest
Bordeaux						
Calais	696					
Lyon	426	612				
Marseille	498	879	270			
Paris	492	228	390	651		
Brest	489	534	759	939	486	

Distances in kilometres taken along straight lines between towns.

Relative directions: Calais, Lyon, Marseille

Calais–Lyon 612 km
Lyon–Marseille 270 km
Total **882 km**
Directly:
Calais–Marseille 879 km

Calais, Lyon and Marseille must lie nearly on a straight line

so that you can tell the difference between this
and the simple distance), and similarly for b and
c, then $a + b = c$ does make sense: it means if I
travel first from Bordeaux to Lyon and then from
Lyon to Calais, I will end up in the same place
as if I simply went from Bordeaux to Calais.
This is how vectors add together. I have added
the displacement from Bordeaux to Lyon to the
displacement from Lyon to Calais to get the
displacement from Bordeaux to Calais.

Bordeaux is placed where the two circular arcs
centred on Lyon and Marseille met. But a check
on that can be made, because we also know the
distance from Bordeaux to Calais (see figure
"Relative directions: locating Bordeaux").

Now that you have your map you can journey
with confidence from one place to another – but
wait a moment! You don't know which way up
the map goes! Which way is north? In fact, the

Relative directions: angled paths

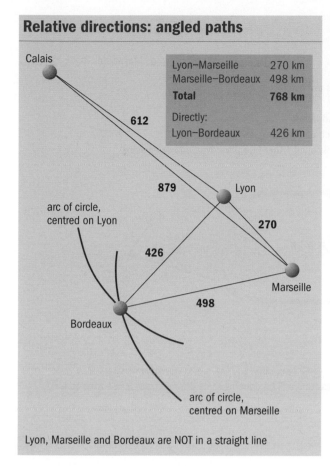

Lyon–Marseille	270 km
Marseille–Bordeaux	498 km
Total	**768 km**
Directly:	
Lyon–Bordeaux	426 km

Lyon, Marseille and Bordeaux are NOT in a straight line

Relative directions: locating Bordeaux

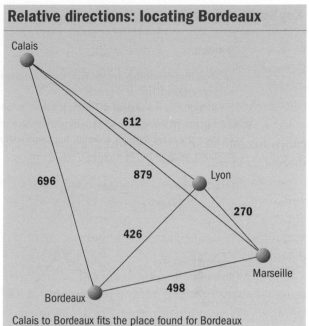

Calais to Bordeaux fits the place found for Bordeaux

Resolving vectors

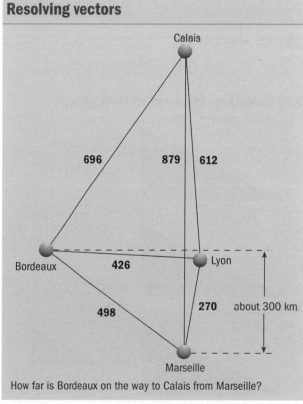

How far is Bordeaux on the way to Calais from Marseille?

relative positions of the towns are independent of which way up the map is: printing maps with north at the top is simply a convention. Notice that vector equations like $a + b = c$ do not depend on the direction of north either. Choosing a special direction is important when dealing with vector components, which you will read about next.

Resolving vectors

If I drive from Marseille to Bordeaux, how far towards Calais have I travelled? Since I was not heading the right way, I have gone 498 km and have 696 km left to go in a different direction.

But, another answer is that I have driven about 300 km along the true direction Marseille–Calais (i.e. close to Lyon). That does not sound very sensible, since at Bordeaux I am nowhere near Lyon, having gone more than 400 km west as well. It sounds a bit more sensible if we ask how far north I have gone from Marseille. Then (if Calais is due north of Marseille, which is roughly true) the answer is about 300 km due north. You can see this in the figure "Resolving vectors", where

the map has been rotated so that it is aligned conventionally, with north at the top.

The distance I have journeyed north is a component of the displacement vector from Marseille to Bordeaux. The displacement from Marseille to Bordeaux can be resolved into two

Key summary: vectors

Notation

To show that quantities are vectors, three notations are used:

a – bold typeface, usual in printed text;

a͟ – straight or wavy underline – the one to use in handwriting;

$\vec{a}$ – arrow over the letter – often used to indicate a vector from A to B, as $\vec{AB}$

To show the magnitude of a vector, the same letter is used either in a plain typeface or by using modulus signs: a or $|a|$ or $|\underline{a}|$ or $|\vec{a}|$.

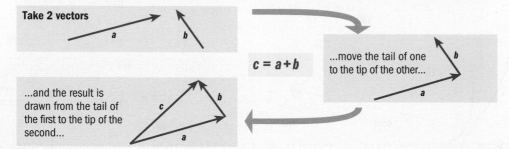

Take 2 vectors

$c = a + b$

...move the tail of one to the tip of the other...

...and the result is drawn from the tail of the first to the tip of the second...

Vector addition

Vectors add by placing them "tip to tail" and drawing the *resultant* vector from the tail of the first to the tip of the second. The *resultant* is the sum of the two vectors. In practice, finding a resultant can be done by scale drawing (as it is in a road atlas!)

Vectors add tip to tail

Key summary: theorem of Pythagoras

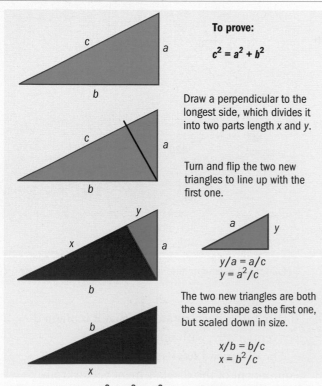

To prove:

$$c^2 = a^2 + b^2$$

Draw a perpendicular to the longest side, which divides it into two parts length *x* and *y*.

Turn and flip the two new triangles to line up with the first one.

$y/a = a/c$
$y = a^2/c$

The two new triangles are both the same shape as the first one, but scaled down in size.

$x/b = b/c$
$x = b^2/c$

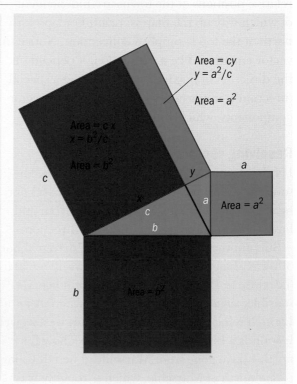

Area = *cy*
$y = a^2/c$

Area = a^2

Area = *cx*
$x = b^2/c$

Area = b^2

Area = a^2

Area = b^2

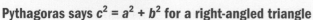

Pythagoras says $c^2 = a^2 + b^2$ for a right-angled triangle

Key summary: components of a vector

red torches cast shadows
of a spider as it walks
along its web in the air

components of spider's path:

vertical
component

4 units horizontally
9 units vertically

horizontal
component

path of
spider
across room

projection of
spider's path
on the wall

projection of
spider's path
on the floor

vector $a = \begin{pmatrix} 4 \\ 9 \end{pmatrix}$

a

9

4

A room provides a co-ordinate system

Key summary: components of a vector from angles

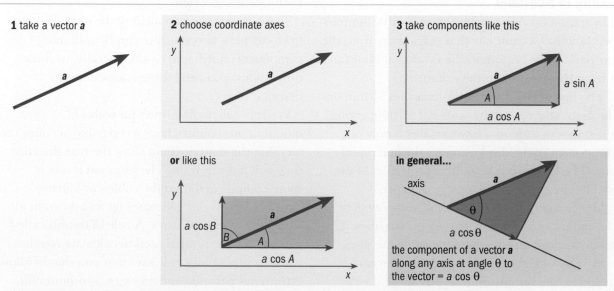

1 take a vector a

a

2 choose coordinate axes

y

a

x

3 take components like this

y

a

$a \sin A$

A

$a \cos A$

x

or like this

y

$a \cos B$

a

B

A

$a \cos A$

x

in general...

axis

a

θ

$a \cos \theta$

the component of a vector a
along any axis at angle θ to
the vector = $a \cos \theta$

The components of a vector depend on the choice of co-ordinates; the vector itself does not

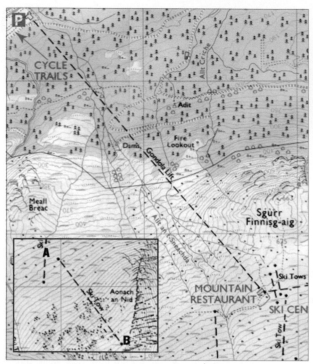

An OS map of Aonach Mor near Fort William in Scotland.

Projection from mountain to flat plane

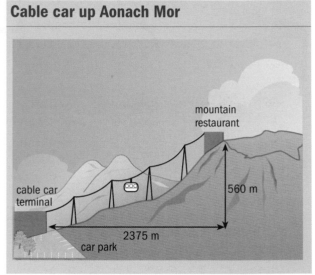

How a 3D mountain is projected onto a flat sheet of paper as a 2D image

Cable car up Aonach Mor

components: 300 km north and 400 km west (both distances are approximate).

But how can you find the components of a vector in given directions? One way of doing this is as we have in "Resolving vectors" p177. We projected the journey from Marseille to Bordeaux onto the direction from Marseille to Calais. You can do this by scale drawing, or by using trigonometry.

Skiing in Scotland

The mountain Aonach Mor near Fort William in Scotland has a cable car that takes skiers from the car park at the bottom to the ski slopes. How far is it from the car park to the restaurant?

The line of the gondola lift measures 95 mm on the map (above) and the scale is 1:25 000, giving a distance of 2375 m. However, this is only one component of the displacement: the horizontal one. The OS map is actually a projection of the real landscape onto a flat piece of paper.

You can get the other, vertical, component of the cable-car journey from the contour lines. The mountain restaurant where the cable car stops is built at a height of 660 m, and the car park is at 100 m: the restaurant is 560 m above the car park. So, to find the straight line distance from the car park to the restaurant, you have to use

Pythagoras's theorem.

The horizontal component of the displacement from car park to restaurant can be split into components north and east. Generally, in three dimensions, you need three components to describe vectors.

Walkers calculating times for walks in mountainous country have to take into account the fact that the map does not show the true distance they will have to walk. The situation is much more complicated than the cable-car journey because real countryside goes up and down in all sorts of complicated ways. A rule of thumb called Naismith's rule is often used to calculate rough walking times. This rule says that you should allow 20 minutes per mile and an extra 30 minutes for every 1000 feet of elevation – or 12 minutes per kilometre plus 10 minutes per 100 m of elevation.

Quick check

1. Show, using Pythagoras's theorem, that I will be 50 miles from my starting point if I first drive 30 miles west and then 40 miles north.

2. Draw a scale drawing to show my total displacement if I first walk 200 m north-west and then 300 m south. Use a scale of 1 cm to 20 m. Show that I will be 212 m from my starting point on a bearing of 222° (i.e. the angle clockwise from north).

3. An aeroplane flies 200 miles south-east. Show by a scale drawing with a scale of 1 cm to 20 miles that the aeroplane has travelled 141 miles south and 141 miles east of its starting point.

4. Show using the map on p180 that
(a) the vertical displacement from A to B is 300 m;
(b) the northwards and eastwards horizontal components are −550 m north and 425 m east. Show that as the ground lies, A and B are about 750 m apart.

5. Fort William in Scotland is just about 100 miles north-west of Edinburgh. Show using trigonometry that Fort William is both 71 miles north and 71 miles west of Edinburgh.

6. Use Pythagoras's theorem to check your answers to question 5 by verifying that the components give the correct distance from Edinburgh to Fort William.

Links to the *Advancing Physics* CD-ROM

Practise with these questions:
60S Short answer *Adding vectors*
30X Explanation–exposition *Constructing a map of France*
40D Data handling *Earthquakes in the USA*
60D Data handling *East Side story*

Try out these activities:
120S Software-based *Tip to tail addition of vectors*
140S Software-based *Understanding vector components*

Look up these key terms in the A–Z:
Displacement; vectors; vector components; vector addition

Go further for interest by looking at:
50T Text to read *Adding vectors*

Revise using the revision checklist and:
50O OHT *Components of a vector by projection*
60O OHT *Components of a vector from angles*

8.3 Velocity

The only vector we have discussed so far is displacement, distance in a certain direction. But if you set off on a journey, you want to know how long it is going to take. The vector related to speed in the same way that displacement is related to distance is **velocity**. Velocity means speed in a certain direction. To calculate velocity, you use:

$$\text{velocity} = \frac{\text{displacement}}{\text{time}} \text{ or in symbols } \boldsymbol{v} = \frac{\boldsymbol{s}}{t}$$

You have to make sure that both sides of the equation are vectors. Velocity and displacement are represented by bold letters, $\boldsymbol{v}$ and $\boldsymbol{s}$, but time, a scalar, is represented by a light typeface t. Note that dividing a vector by a scalar on the right hand side of the equation gives you a vector.

Just as we can have an average speed and an instantaneous speed, so we can talk about average velocity and instantaneous velocity. While reading this book, you are on a journey, moving at a speed of about 30 kilometres per second. Why? Because the Earth is travelling round the Sun (note that we have ignored the spin of the Earth about its axis, the rotation of the galaxy and so on).

What about your velocity? We don't know what that is. Why not? Because to tell you your velocity, we have to be able to give you a direction as well as a speed and because we don't know when you are reading this, we don't know the whereabouts of the Earth in its orbit. Are your average and instantaneous velocities the same? No. The direction in which you are moving is changing second by second as the Earth moves in its orbit. Notice how careful you have to be when you talk about vectors: here, there is a constant speed, a velocity that changes second by second and an average velocity that is equal to zero.

Adding velocities

If you have ever flown across the Atlantic, you will know that the time to fly from west to east is less than the time it takes to fly from east to west. This is because of the jet stream, a current of air that blows more or less steadily across the Atlantic from west to east. It is produced by convection currents in the atmosphere, between the warm tropics and the cold north polar region. The aeroplane

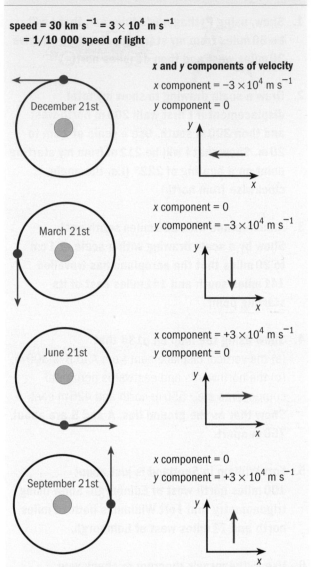

Key summary: speed, velocity of Earth's orbit

speed = 30 km s^{-1} = 3 × 10^4 m s^{-1}
= 1/10 000 speed of light

x and y components of velocity

December 21st
x component = −3 × 10^4 m s^{-1}
y component = 0

March 21st
x component = 0
y component = −3 × 10^4 m s^{-1}

June 21st
x component = +3 × 10^4 m s^{-1}
y component = 0

September 21st
x component = 0
y component = +3 × 10^4 m s^{-1}

For the Earth going round the Sun: speed is large and constant; velocity changes all the time; and average velocity over a year is zero

flies through the air at the same speed in both directions, but not at the same speed over the ground. This is because the air is moving too. To find the plane's velocity relative to the ground we have to add together the velocity of the air and the velocity of the aeroplane.

Suppose now that as the pilot flies from Kennedy Airport, New York, to Heathrow, London, the plane has to cope with a wind blowing from north to south across its flight path. What happens if the pilot simply ignores this and tries to fly on a line

Speeds

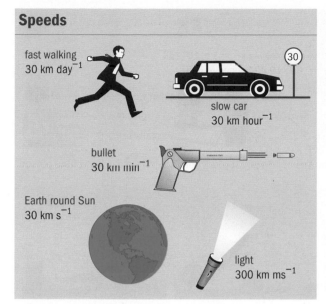

fast walking
30 km day^{-1}

slow car
30 km hour^{-1}

bullet
30 km min^{-1}

Earth round Sun
30 km s^{-1}

light
300 km ms^{-1}

Vector addition

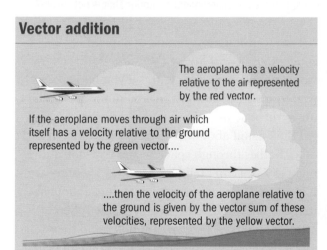

The aeroplane has a velocity relative to the air represented by the red vector.

If the aeroplane moves through air which itself has a velocity relative to the ground represented by the green vector....

....then the velocity of the aeroplane relative to the ground is given by the vector sum of these velocities, represented by the yellow vector.

Aeroplane, velocity vectors and calculations

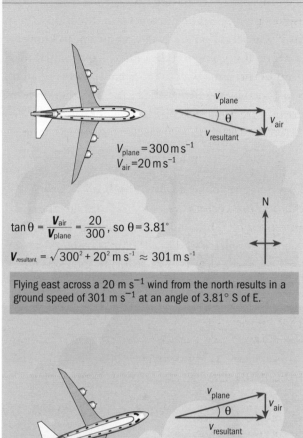

$V_{plane} = 300 \, m \, s^{-1}$
$V_{air} = 20 \, m \, s^{-1}$

$\tan \theta = \dfrac{V_{air}}{V_{plane}} = \dfrac{20}{300}$, so $\theta = 3.81°$

$V_{resultant} = \sqrt{300^2 + 20^2} \, m \, s^{-1} \approx 301 \, m \, s^{-1}$

Flying east across a 20 m s^{-1} wind from the north results in a ground speed of 301 m s^{-1} at an angle of 3.81° S of E.

$V_{plane} = 300 \, m \, s^{-1}$
$V_{air} = 20 \, m \, s^{-1}$

$\sin \theta = \dfrac{V_{air}}{V_{plane}} = \dfrac{20}{300}$, so $\theta = 3.82°$

$V_{resultant} = \sqrt{300^2 - 20^2} \, m \, s^{-1} \approx 299 \, m \, s^{-1}$

In order to fly due E, the plane has to fly 3.82° N of E. The resultant ground speed is 299 m s^{-1}.

Vector addition allows airline pilots to calculate their speed relative to the ground.

west to east? Adding the plane's velocity vector to that of the wind, the plane flies, relative to the ground, in the wrong direction! To find out which way it is flying, you can use trigonometry.

Which way should the pilot fly the plane, relative to the air, in order to travel west to east relative to the ground? The plane can only fly at a speed of 300 m s^{-1} relative to the air, so it has to fly at an angle so that the combined effect of the wind velocity and the aeroplane's velocity is in the right direction.

Components of velocity

Just as you can think about the components of a displacement, velocity too has components. Imagine taking a video recording of someone walking diagonally across the room in front

of you. What do you see when you watch your film-making efforts? On the TV screen, you see someone moving across the screen from right to left: you don't see the forward motion of the walker except as an apparent increase in their size. What you see is only one component of their motion, and if you calculate their speed from right to left, you have, in fact, calculated only one component of their velocity. Their actual motion is projected onto a flat screen just as the OS map projects

These video stills show only the motion perpendicular to the camera's line of sight, if you ignore the increase in size of the person.

The video stills of a falling chimney show vertical and horizontal motion. Is there be a component of motion that is not visible?

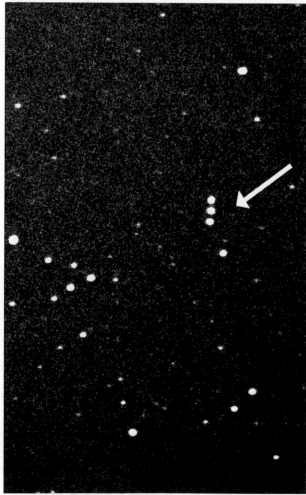

A visibly moving star (Barnard's star). Barnard's star (indicated by the arrow) was photographed three times at two year intervals and the images superimposed. It has moved against the background of other stars. Of course only the components of its motion at right angles to the line of sight can be seen.

rugged terrain onto a flat piece of paper. (This is not quite true: if it were, the size of the person would not change.) Of course, if you had taken your video from above, you would have been able to see the true motion.

Think again about the journey of the Earth and other bodies through space: when you look at the night sky you can't tell how far away the stars are. It is easy to understand why it was once thought that the stars were holes in a heavenly sphere.

But, of course, stars are not holes and they move through space. Every night, you can watch the constellations move round the sky because of the rotation of the Earth, but over much longer periods of time it is possible to detect that some stars have moved relative to their neighbours. The motion seen is only one component of the motion of the stars. You cannot see the motion of stars towards or away from you, although this motion can in fact be detected by changes in spectral lines. Only the projection of the motion onto the heavenly sphere, the two components of the motion at right angles to your line of sight, is visible as a change in position.

Vectors everywhere

You will find vector quantities appearing again and again in physics and engineering, from the design of bridges to the study of ocean currents and the motion of galaxies in the universe.

Examples of vector quantities

Ocean currents in the Atlantic

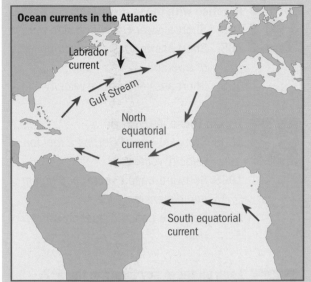

Labrador current

Gulf Stream

North equatorial current

South equatorial current

Winds in the atmosphere

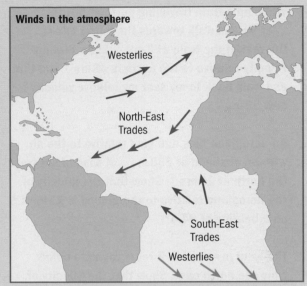

Westerlies

North-East Trades

South-East Trades

Westerlies

Forces in a bridge

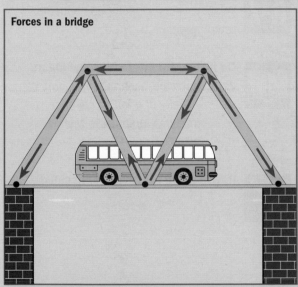

Surveying lines

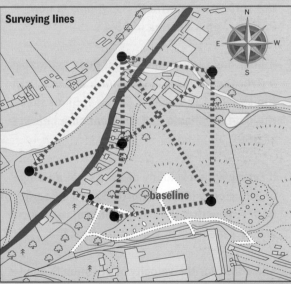

baseline

Field near a magnet

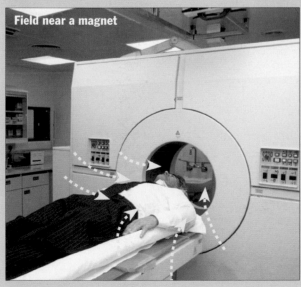

Air flow over a car

K-LN-000

Quick check

1. While on a train travelling north at $50\,\text{m s}^{-1}$, I get up and walk towards the buffet car at the rear of the train at $2\,\text{m s}^{-1}$. Show that my velocity relative to the Earth is $48\,\text{m s}^{-1}$ and that as I walk back to my seat my relative velocity is now $52\,\text{m s}^{-1}$.

2. An aeroplane flies due east relative to the air with an airspeed of $250\,\text{m s}^{-1}$. A wind blows from the north at $20\,\text{m s}^{-1}$. Show that the velocity of the aeroplane relative to the ground is $251\,\text{m s}^{-1}$ on a bearing of 94.5°.

3. The wind in question 2 veers round to come from the north-east. Show that the velocity of the aeroplane relative to the ground is now $236\,\text{m s}^{-1}$ on a bearing of 93.4°.

4. A boat has to be made to sail due east across a river that is flowing due south at $5\,\text{m s}^{-1}$. The top speed of the boat is $10\,\text{m s}^{-1}$. Show that the captain should sail the boat on a bearing of 060°.

5. If I drive with a velocity of $20\,\text{m s}^{-1}$ at an angle of 30° clockwise from north (i.e. on a bearing of 030°) show that the components of my velocity are $17.3\,\text{m s}^{-1}$ north and $10\,\text{m s}^{-1}$ east.

6. Show that for the velocity of question 5, the component of my velocity north-west is $5.2\,\text{m s}^{-1}$ and $19.3\,\text{m s}^{-1}$ north-eastwards.

Links to the *Advancing Physics* CD-ROM

Practise with these questions:
70S Short answer *Flying in a side wind*
80S Short answer *Using the components of a vector*
90S Short answer *Relative velocity*

Try out these activities:
180S Software-based *Getting a feel for velocity vectors*
190S Software-based *Vector addition in one dimension*
200S Software-based *A boat crossing a river*

Look up these key terms in the A–Z:
Velocity; vectors; vector components; vector addition

Go further for interest by looking at:
220S Software-based *Components of velocity of a falling chimney* (requires Multimedia Motion CD-ROM)

Revise using the revision checklist

Summary check-up

Motion ✓

- Speed $= \dfrac{\text{distance travelled}}{\text{time taken}}$; $v = \dfrac{s}{t}$

- The gradient of a distance–time graph gives speed
- The area under a speed–time graph gives distance travelled
- Distance–time and speed–time graphs can give alternative views of the same journey
- Both distance–time and speed–time graphs can be interpreted to give a description of a journey
- The average speed is the total distance divided by the total time
- The instantaneous speed is the speed at a given moment, equal to the gradient of a distance–time graph

Vectors ✓

- Scalar quantities have no direction; vector quantities are fully specified only if a magnitude and a direction are given (or two components)
- Distance and speed are scalar quantities, displacement and velocity are the corresponding vectors
- Vectors add geometrically, that is by placing them tip to tail, to form a resultant vector
- Vectors can be added algebraically; special cases are: the resultant is the sum of the magnitudes if the vectors are in the same direction; the resultant is the difference between the magnitudes if the vectors are in opposite directions; the resultant is given by Pythagoras's theorem if the vectors are at right angles
- The length of a vector is independent of a co-ordinate system but its components are not
- Vectors may be resolved into components at right angles

Questions

1. **Below is a distance–time graph showing the height of a stone dropped from a tall building.**

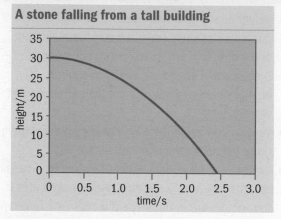

A stone falling from a tall building

(a) What is the height of the building?
(b) At what speed was the stone dropped? Explain your answer.
(c) Use the graph to estimate the speed of the stone as it hits the ground.
(d) Use the graph to calculate the average speed of the stone during its fall. How does this compare with (i) the speed of the stone after half the time between release and hitting the ground has passed; (ii) the speed of the stone after the stone has fallen half the height of the building; (iii) the average of the speed at the top and the speed at the bottom of the fall.
(e) Draw a speed–time graph of the fall of the stone.

2. **Table 1 gives distances "as the crow flies" in kilometres between a cluster of towns.**

Table 1

	Leeds	Huddersfield	Batley	Bradford	Bingley
Huddersfield	22.5				
Batley	11.8	10.7			
Bradford	13.2	16.2	10.5		
Bingley	19.7	22.2	18.5	8.2	
Wakefield	13.0	19.2	12.0	20.7	28.7

(a) Without making a drawing, give a reason to think that Leeds, Batley and Huddersfield lie along a straight line.
(b) Still without drawing, give a reason to think that Leeds, Bradford and Huddersfield do not lie on a straight line.
(c) Use the distances to construct a possible map of the layout of these towns.
(d) Turn or flip your map to produce an alternative map. What does not change as you do so?

3. **The graph shows part of the journey of an intercity train.**

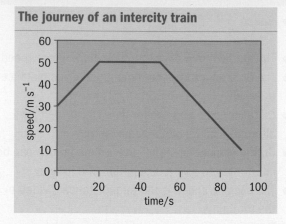

The journey of an intercity train

(a) Write a few sentences to describe the journey. In what ways is the graph unrealistic?
(b) How far did the train travel during the first 20 s shown on the graph?
(c) How far did the train travel during the 90 s shown on the graph?
(d) Draw a distance–time graph of the journey.

4. **A girl can swim at 1.2 m s⁻¹. She swims across a river flowing at 0.5 m s⁻¹.**
(a) If she swims perpendicularly to the flow of the water, what is her velocity relative to the bank?
(b) If the river is 10 m wide, how long would it take her to swim across the river?
(c) In what direction relative to the water would she have to swim to move perpendicularly to the bank?
(d) In the situation described in part (c), how long would it take her to swim across the river?

5. **The data in table 2 were taken for a car.**

Table 2

Time/s	0	1	2	3	4	5	6	7	8	9
Speed/m s⁻¹	0	3.2	6.4	9.4	12.5	15.4	18	20	20.8	20.8

(a) Draw a speed–time graph for these data.
(b) Describe the car's motion during the first 9 s.
(c) Suggest reasons why the car reached a steady speed.
(d) Use the graph to estimate how far the car travelled (i) in the first 5 s; (ii) in the first 9 s.

9 Computing the next move

As predicted, after a journey of several years, a spacecraft arrives at a planet and sends back pictures. Your holiday aircraft approaches its destination: "10 minutes to landing" the captain announces. We will tell the story of how movements are computed ahead of time and show you:

- how effects of relative movement can be anticipated
- how to calculate speeds, distances and times when things accelerate
- how to compute the path of an object acted on by gravity
- how to calculate changes of energy of moving objects

9.1 What's the next move?

You exist in a world of movement – and you live on a moving world. At the end of the 20th century more people are moving faster and farther than ever before. Or perhaps not so fast – the average speed of motor traffic in central London is almost exactly what it was for horse-drawn vehicles in 1900. On the other hand, people pop across to New York for the weekend and more than 100 000 people cross the English Channel every day.

The fastest and most agile military aircraft are controlled by computer because, despite rigorous training, human pilots do not have rapid enough reactions. The computers sense what the aircraft is doing and act accordingly. Computing the next move is an elegant and powerful part of physics.

But not all movement is to do with traffic. Besides walking and running, many people get pleasure from games demanding the skilful control or prediction of the movement of a ball.

How you catch a ball

The game is on – the ball flies through the air towards you. You have to get your hand or racquet

A tennis player, constantly keeping his eye on the ball, computes the optimum moment to make contact for the shot.

A bat catches a moth using ultrasound ranging to keep track of where it is. Somehow the bat has to work out what to do next.

A modern aircraft, such as the Eurofighter Typhoon in simulation here, provides a pilot with a wealth of information but its flight controls are so finely tuned that the human brain and body are simply too slow to cope.

Key summary: tracking a moving object

aircraft
approaching airport

radar at
airport

distance 10 km

speed 360 km per hour
= 100 m s^{-1}

Measure and predict at short time intervals

Aircraft reduces speed and turns.
Need frequently updated velocity measurements.

Aircraft travels at constant speed in same direction so
velocity is constant and you can predict well ahead.

displacement in a short
time interval

displacement Δs in time Δt is $v\Delta t$

example: velocity v = 100 m s^{-1}
displacement in 10 seconds = 1000 m

Displacement $\Delta s = v\Delta t$

How you catch a ball

To make a catch, keep your eye on the ball. Run at just the
right speed to keep the angle of your line of sight constant. As
the ball gets closer to you this becomes more difficult. You will
have to adjust your speed or even move backwards a little to
get it right. Often young children who have not yet learned to
make that last-minute adjustment get hit in the eye. The trick
is always to keep the vector of the relative motion between you
and the ball in the same direction.

in the right place to meet the ball at the right time.
How is it done? Most skilful players don't know;
they just do it. Practice has taught their minds and
muscles how to react faster than they can think. If
you are lucky, you manage to get your hand to be
in the same place as the ball at the same time. The
catch is made, the game is won.

A little thinking reveals the hidden rule that
your brain and body has learned to use. You keep
your eye on the ball. That is, you sense the angle
of the direction from eye to ball and then run so
as to maintain that angle as the ball moves. That's
all. If you can move so as keep the angle to the
ball constant then you and it are on a collision

course, and you'll catch it (in the eye, if you don't
make a last-minute adjustment!). The skill comes
in rapidly altering your running, to keep the ball
seeming to come from the same direction. Watch
players doing it, or try it yourself, and you'll see
how it works. But when you want to win, don't slow
yourself down by thinking. Just do it.

Air-traffic control

Every day, more than 1000 aircraft enter the air
space over London's Heathrow Airport. Each
aircraft is identified, and its height, distance,
direction and velocity are monitored by radar. Air-
traffic control uses a large and complex computer
system to handle all these data – the primary aim
is to predict and act in time to avoid collisions.

You may have seen the radar aerial at an airport
turning round and round, surveying the sky for
aircraft. Having found an aircraft and measured
its distance and velocity, the aerial has to turn
right round again before there is another chance
to make measurements. So data only come in at
intervals. But if the intervals are short it is safe to
predict the next position of the aircraft, because
the velocity cannot change much in a short time.
The controllers can imagine the flight path as a
sequence of short straight-line steps.

At velocity v, a step would be a displacement of
$v\Delta t$ in time interval Δt. This may be farther than

you think. An aircraft coming in at 360 km per hour, much less than its typical mid-flight speed, still travels at 100 m s^{-1}. That's one kilometre every 10 seconds. Typically the radar dish rotates once every few seconds, say about five seconds.

If air-traffic control knows that the aircraft is not going to change speed or turn, the future position can be computed much further ahead. The path is a straight line, moving by equal displacements in equal time intervals. It isn't safe to predict very far ahead like this though, because this involves the risky belief that the future will be like the past. If the aircraft is changing speed or turning, computing the next move is more work.

If the velocity (speed and direction) can be measured at each time interval, air-traffic control can look ahead in time just by that time interval. Of course, if there is a flight plan, telling the expected speeds and directions at frequent intervals, the prediction can be made further ahead, plotting a chain of displacement vectors, one for each time interval. But that involves the risky assumption that things will go according to plan, and remember that the displacements could be as much as a kilometre long.

Warning! Collision course

Air-traffic control has to anticipate possible air collisions in time to tell the pilots to change direction, and time may be short. Modern planes travel at around 1000 km per hour. Two such planes travelling directly towards one another are coming together at a relative velocity of 2000 km per hour. When 10 km apart there is less than 20 seconds in which to act. That is why computers are needed to look into the future for each aircraft to detect danger as quickly as possible. In fact, air-traffic-control rules require aircraft on opposing paths to change course before they are about 60 km apart.

Generally, aircraft will approach each other at an angle and with different velocities relative to the ground. Then the question is, will they collide? Predicting a collision is predicting something that you want to prevent. So you look ahead, asking a "what if" question: what will happen if the two aircraft go on flying at their present velocity? This requires a calculation that looks into the future.

How is it done? One way is to plot out the two

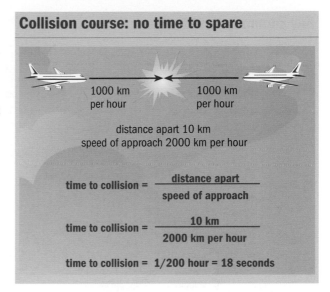

Collision course: no time to spare

1000 km per hour   1000 km per hour

distance apart 10 km
speed of approach 2000 km per hour

$$\text{time to collision} = \frac{\text{distance apart}}{\text{speed of approach}}$$

$$\text{time to collision} = \frac{10 \text{ km}}{2000 \text{ km per hour}}$$

$$\text{time to collision} = 1/200 \text{ hour} = 18 \text{ seconds}$$

paths step by step and see if they arrive somewhere at the same moment. Another is to use the idea of relative motion.

Relativity at low velocities

You are a passenger in an aircraft. You look out of the window. The Sun glints off the wings of another aircraft far away to the left. A few moments later the other plane is closer and you get out your camera to take a photo of it. By the time you have got it in focus it is much nearer and you feel a twinge of alarm. Then you start to get really worried as the other aircraft gets larger and larger in your viewfinder, but – just in time – it zooms ahead and passes in front of your plane.

From your point of view the other aircraft is coming towards you at a fixed angle and speed. Sitting in your plane, you see things as if you were not moving. The movement that you see of the other plane is in fact the relative velocity of the two aircraft. Passengers in the other aircraft would see your plane coming towards them.

Both aircraft are moving relative to the ground. In the vector diagram on p192, your plane has been "stopped" by adding an equal but opposite velocity to its velocity. To put it another way, your plane's velocity has been subtracted from itself to give zero. You can't really stop the plane, of course. You must subtract the same velocity from everything else, including the ground. So with your plane "stopped", the ground is now slipping away behind you at that velocity, which is exactly

Key summary: an air miss

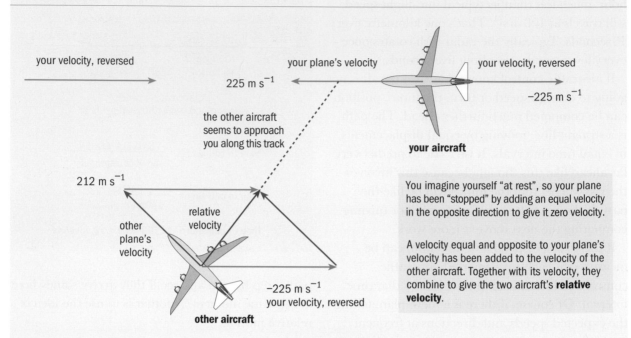

your velocity, reversed

your plane's velocity
225 m s^{-1}

your velocity, reversed
-225 m s^{-1}

the other aircraft seems to approach you along this track

your aircraft

212 m s^{-1}

relative velocity

other plane's velocity

-225 m s^{-1}
your velocity, reversed

other aircraft

You imagine yourself "at rest", so your plane has been "stopped" by adding an equal velocity in the opposite direction to give it zero velocity.

A velocity equal and opposite to your plane's velocity has been added to the velocity of the other aircraft. Together with its velocity, they combine to give the two aircraft's **relative velocity**.

Rules:
1. **Add a velocity opposite to that of one plane to the velocities of both**
2. **Find the resultant relative velocity, adding vectors tip to tail**
3. **See if the direction of the relative velocity hits your plane. If so, take avoiding action!**

what you see as you look out of the window, thinking of yourself as not moving.

What about the other aircraft? Is it now coming directly towards you? To find out, you subtract your plane's velocity from the other aircraft's velocity too. That is, you find the vector sum of the second plane's velocity relative to the ground and the reverse of your velocity relative to the ground. Since your plane has been "stopped", if the velocity of the other plane points directly towards you, there is going to be a collision unless something is done about it. The resultant vector shows the **relative velocity** of the other aircraft with respect to yours.

Ships' captains know that there is danger of a collision if the compass bearing of another ship as seen from their ship stays the same. This means that their relative velocity points along the line joining them. It's the same problem as that of catching a ball, except that in the case of the ball you are trying to get on a collision course with it. You run towards the falling ball at just the right

velocity to keep its direction to you constant. You – and airline pilots and sailors – need to keep your eye on the ball.

Bats and astronomers

Like airport radar, a bat catching an insect (p189) measures the distance of a remote object, but by sound ranging rather than radio ranging. Astronomers now measure the size of the solar system by radar. They are interested in the relative velocities of astronomical objects too – of meteorites that might hit Earth, and of distant galaxies.

In these and other examples, distance is measured not with rulers but with clocks. Time – the travel time of a signal – becomes the measure of distance. This is why astronomical distances are often stated in light-years. And because the speed of light is a fundamental constant, it gives a way of making a fundamental link between space and time. Also, it means that distant objects are seen as they were at an earlier time – looking out into space is in fact looking back in time.

Quick check

1. Your car is going at $25\,\text{m s}^{-1}$ along a straight narrow road. Another car comes the other way at $35\,\text{m s}^{-1}$. Show that the relative speed of the two cars is $60\,\text{m s}^{-1}$.

2. Your car heads north at $25\,\text{m s}^{-1}$. Another car approaches along a side road, going west at $20\,\text{m s}^{-1}$. Both are $100\,\text{m}$ from where the roads join. Show that the velocity of the other car relative to your car is $32\,\text{m s}^{-1}$ at an angle of $39°$ to the main road.

3. A canoeist points her canoe directly across a $100\,\text{m}$ wide river. She paddles at $2\,\text{m s}^{-1}$; the river flows at $1.5\,\text{m s}^{-1}$. Show that it takes $50\,\text{s}$ to cross the river and that the canoe will travel $75\,\text{m}$ downstream in this time.

4. Next time, she does a "ferry glide". She points the canoe upstream so that the resultant motion is directly across the river. Sketch velocity vectors to show that this is possible.

5. Using a scale diagram or by calculation, show that the resultant speed of the canoe using the ferry glide is about $1.3\,\text{m s}^{-1}$.

6. Show that it takes $77\,\text{s}$ to cross the river using the ferry glide.

7. Air-traffic control detects an aircraft on a bearing $090°$ (due east). Radar pulses return with a delay of $0.2\,\text{ms}$. $100\,\text{s}$ later it is detected on a bearing of $120°$, with a pulse delay of $0.173\,\text{ms}$. Show that the velocity of the aircraft is $150\,\text{m s}^{-1}$ on a bearing of $210°$.

Links to the *Advancing Physics* CD-ROM

Practise with these questions:

10M   Multiple choice *Displacement–time graph*

20S   Short answer *Calculated steps*

Try out these activities:

10S   Software-based *Reconstructing a flight*

30S   Software-based *Will the aircraft collide?*

Look up these key terms in the A–Z:

Relative velocity; vector addition; velocity

Go further for interest by looking at:

10S   Computer screen *Multiflash photographs*

Revise using the revision checklist and:

20O   OHT *Relative velocity*

9.2 Speeding up, slowing down

A sprinter obviously wants to reach their top speed as soon as possible after the start of a race. Acceleration is the rate of change of velocity. Many accelerations and decelerations go unnoticed around you. For example, as you read this page, your eyes make tiny jumps across the words. To do that, muscles must set your eyes rotating and then stop them again. If you move a hand you have to give it a velocity. If you hit a ball high in the air it slows down going up and then speeds up coming down.

A golf ball accelerates when struck by a golf club.

Most of these accelerations are over quickly, but to think about acceleration it is useful to think about what happens if it continues at a constant rate – if the same amount of velocity is added in the same interval of time. This is the case of **uniform acceleration**. To keep things simple, it's best to begin with acceleration in a straight line.

If you know that the acceleration is constant, you can predict ahead of time where something will get to and how fast it will be moving. Or, as in the case of preventing an air collision (p192), you can make a "what if" prediction: what would happen if the acceleration were constant.

A simple way to make predictions is to draw arrows for the velocity and displacement vectors. Divide time into small slices and in each small slice of time add a little to the velocity vector. If the acceleration is uniform, the amount added is the same each time. Each new velocity vector is

Left: the vaulter accelerates as their pole straightens. Right: even when moving upwards a diver is being accelerated downwards.

the one from the interval before, with a small fixed extra length of arrow attached, as the object goes faster and faster. The displacement in each time interval is then another vector equal to the velocity at that time multiplied by the time interval.

Key summary: uniform acceleration

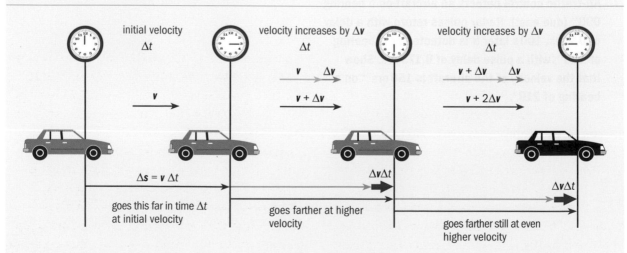

A uniformly accelerating object adds the same amount to its velocity in each short time interval.

Key summary: computing rules for accelerated motion

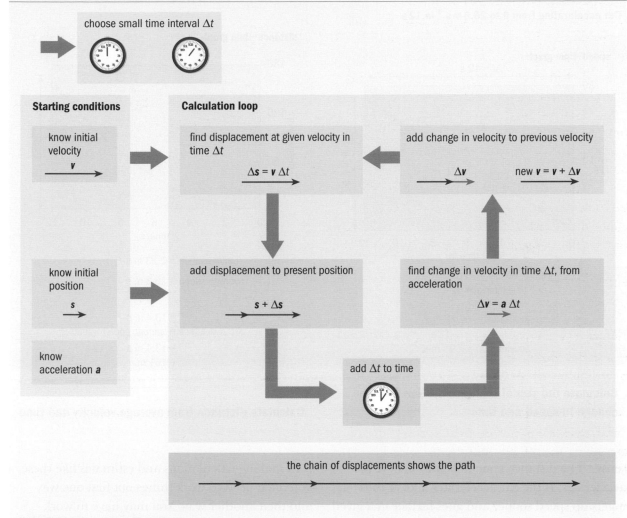

choose small time interval Δt

Starting conditions

know initial velocity

v

know initial position

s

know acceleration a

Calculation loop

find displacement at given velocity in time Δt

$$\Delta s = v \, \Delta t$$

add displacement to present position

$$s + \Delta s$$

add Δt to time

find change in velocity in time Δt, from acceleration

$$\Delta v = a \, \Delta t$$

add change in velocity to previous velocity

Δv new $v = v + \Delta v$

the chain of displacements shows the path

Calculate in a repeating loop:
1. Get change in displacement from current velocity; add to present displacement
2. Get change in velocity from acceleration; add to present velocity
3. Go to 1 with new time $t + \Delta t$

Reconstructing a traffic incident

An impatient driver is waiting at traffic lights. They go green and he accelerates away. Police video and radar show that 12 s later the car was going at 60 mph past a shop later measured to be 160 m from the lights. The driver claims that the shop is too near the lights for that speed to be possible. Who's right? The incident needs to be reconstructed.

First check that the claimed acceleration is believable. Many cars can accelerate faster than 0–60 mph in 12 s. The acceleration was at the rate of 5 mph in each second, on average.

For a different check, convert the acceleration into SI units. 60 mph is 26.8 metres per second (a mile is 1.609 km). Over the 12 s, the car had to gain speed at a rate of $\frac{26.8 \text{ m s}^{-1}}{12 \text{ s}} = 2.23 \text{ m s}^{-1}$ per second, on average. The clumsy unit "metres per second, per second" can be abbreviated to m s^{-2}. This is the SI unit of acceleration. 2.23 m s^{-2} is not an impossible acceleration for a car. It is about one-fifth of the acceleration of a freely falling body, which is about 10 m s^{-2}. This is quite a rapid increase in speed, but not out of the question. It is always useful to cross check a result against some other quanitiy that you know.

Does the claimed distance of 160 m make

Key summary: constant acceleration

Car accelerating from 0 to 26.8 m s^{-1} in 12 s

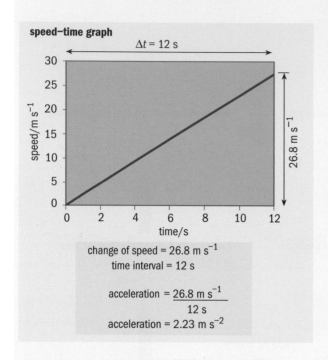

speed–time graph

$\Delta t = 12$ s

change of speed = 26.8 m s^{-1}
time interval = 12 s

acceleration = $\dfrac{26.8\ \text{m s}^{-1}}{12\ \text{s}}$

acceleration = 2.23 m s^{-2}

**Calculate the size of the acceleration from the
change in speed and time**

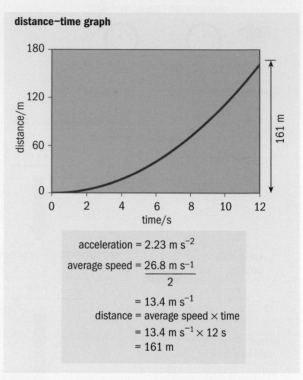

distance–time graph

161 m

acceleration = 2.23 m s^{-2}

average speed = $\dfrac{26.8\ \text{m s}^{-1}}{2}$

= 13.4 m s^{-1}

distance = average speed × time

= 13.4 m s^{-1} × 12 s

= 161 m

Calculate distance from average velocity and time

sense? The distance gone depends on how the car accelerates. If the car accelerates a lot at the start, it gets up speed sooner and goes farther in a given time. What if the acceleration is roughly constant?

If the car did go 160 m in 12 s, its average speed would have been: $\dfrac{160\ \text{m}}{12\ \text{s}} = 13.3\ \text{m s}^{-1}$. This turns out to fit in with the claimed acceleration. Here's how. The car was claimed to go from 0 to 26.8 m s^{-1}, so if its acceleration were constant the average speed would have been: $\dfrac{0\ \text{m s}^{-1} + 26.8\ \text{m s}^{-1}}{2} = 13.4\ \text{m s}^{-1}$. Doing the calculation the other way round, if the average speed had been 13.4 m s^{-1}, lasting for 12 s, the car would have travelled a distance:

$$s = vt$$

$$s = 13.4\ \text{m s}^{-1} \times 12\ \text{s} = 161\ \text{m}$$

The reconstruction all seems to fit together. The various measurements and estimates agree. Check them through for yourself to see how they work.

In making calculations and estimates like these, you often need to work things out first one way and then another way. You may have to work out distance from speed and time, or speed from distance and time.

Logic lets you look ahead

The traffic incident reconstruction used only two simple ideas, true for constant acceleration in a straight line:

$$\text{acceleration} = \frac{\text{change of velocity}}{\text{time}}$$

$$= \frac{\text{final velocity} - \text{initial velocity}}{\text{time}}$$

$$\text{average velocity} = \frac{\text{initial velocity} + \text{final velocity}}{2}$$

We think that one of the pleasures of doing physics is how, sometimes, from such simple and almost obvious starting points, you can work out a

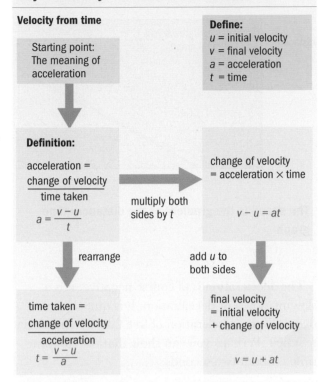

Key summary: uniform acceleration 1

Velocity from time

Starting point: The meaning of acceleration

Define:
u = initial velocity
v = final velocity
a = acceleration
t = time

Definition:

$$acceleration = \frac{change\ of\ velocity}{time\ taken}$$

$$a = \frac{v - u}{t}$$

multiply both sides by t

change of velocity = acceleration × time

$$v - u = at$$

rearrange

time taken = $\frac{change\ of\ velocity}{acceleration}$

$$t = \frac{v - u}{a}$$

add u to both sides

final velocity = initial velocity + change of velocity

$$v = u + at$$

Use $v = u + at$ to compute velocity from acceleration and time

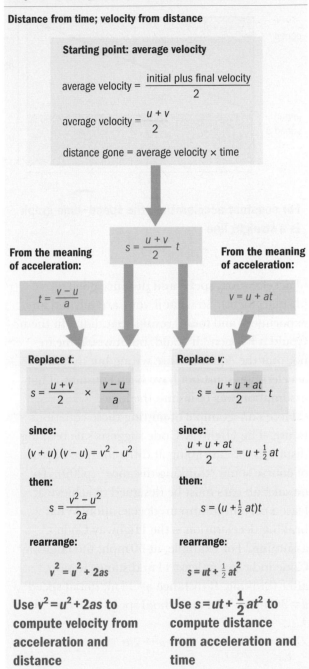

Key summary: uniform acceleration 2

Distance from time; velocity from distance

Starting point: average velocity

$$average\ velocity = \frac{initial\ plus\ final\ velocity}{2}$$

$$average\ velocity = \frac{u + v}{2}$$

distance gone = average velocity × time

$$s = \frac{u + v}{2}\,t$$

From the meaning of acceleration:

$$t = \frac{v - u}{a}$$

From the meaning of acceleration:

$$v = u + at$$

Replace t:

$$s = \frac{u + v}{2} \times \frac{v - u}{a}$$

since:

$$(v + u)(v - u) = v^2 - u^2$$

then:

$$s = \frac{v^2 - u^2}{2a}$$

rearrange:

$$v^2 = u^2 + 2as$$

Use $v^2 = u^2 + 2as$ to compute velocity from acceleration and distance

Replace v:

$$s = \frac{u + u + at}{2}\,t$$

since:

$$\frac{u + u + at}{2} = u + \tfrac{1}{2}at$$

then:

$$s = (u + \tfrac{1}{2}at)t$$

rearrange:

$$s = ut + \tfrac{1}{2}at^2$$

Use $s = ut + \dfrac{1}{2}at^2$ to compute distance from acceleration and time

whole lot of other things. These two simple ideas have several logical consequences, which you might not guess just by looking at them, but with a little work, these consequences can be derived. The consequences are hidden inside the two simple equations for acceleration and average velocity. Having brought them out in the open, they are ready to hand to make all sorts of practical predictions about motion.

Useful kinematic equations

The four most useful relationships, all proved just from the meaning of acceleration and the idea of average speed, are:

$$v = u + at$$

$$s = \frac{u + v}{2}t$$

$$s = ut + \frac{1}{2}at^2$$

$$v^2 = u^2 + 2as$$

Strictly speaking, by "velocity", "acceleration" and "distance gone" here we mean their components along the direction of motion. With extra notation, not used here, the equations can be written in vector form.

Here's a thought. There is no point at all in testing these relationships experimentally. They can't be wrong. If the acceleration is constant, they follow logically just from the meanings

Key summary: graphs of constant acceleration

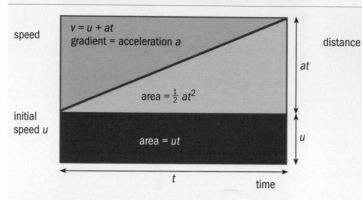

For constant acceleration, the speed–time graph is a straight line

The speed is the gradient of the distance–time graph

of acceleration, speed and distance gone (displacement). So what if you carefully did an experiment and found results that didn't fit them? Could it happen? It could, but it would mean not that the equations are wrong but that the acceleration must not have been constant. That assumption was built into the logic.

Here's an example of putting these equations to use. The Highway Code suggests safe braking distances for cars going at different speeds, plus of course some "thinking distance" (p209). To be safe, all cars must be designed to achieve at least a certain minimum deceleration. What braking deceleration is the Highway Code assuming? For example, at 30 mph, the Highway Code indicates at least 14 m distance to stop, after thinking. If distance $s = 14$ m; initial speed $u = 30$ mph $= 13.4$ m s^{-1}; final speed $v = 0$ m s^{-1}. Use:

$$v^2 = u^2 + 2as$$

You want the value of a, so rearrange the equation:

$$2as = v^2 - u^2$$

$$a = \frac{v^2 - u^2}{2s}$$

Substitute values:

$$a = \frac{(0 \text{ m s}^{-1})^2 - (13.4 \text{ m s}^{-1})^2}{2 \times 14 \text{ m}}$$

$$a = \frac{-180 \text{ m}^2 \text{ s}^{-2}}{2 \times 14 \text{ m}} = -6.4 \text{ m s}^{-2}$$

The acceleration is of course negative: it is a slowing down or deceleration. It is quite large – over half the acceleration of free fall. You need that seat belt. Perhaps you can show that the stopping time is about two seconds.

Modelling and predicting using graphs

Graphs help us to picture motion. They are most useful when the movement is uneven or complicated, when equations are less help. A simple case where acceleration is constant helps you to understand how the shape of a graph is connected to the kind of motion it represents.

The area below a graph of speed against time is the displacement (chapter 8). If the acceleration is constant, this area can be thought of as being in two parts:

- a rectangle of height u, width t, area ut, which is the distance that would have been travelled in time t;
- a triangle, height at, width t, area $\frac{1}{2}at^2$, which is the extra distance travelled because the velocity is increasing.

Often acceleration will not be uniform. For example, a car pulling away from traffic lights might accelerate gently at first, then build up the acceleration, and then finally reduce it as the car gets to the desired speed.

Whatever the shape of the graph:

- the slope at any point on a distance–time graph gives the speed at that point in time;
- the area under a speed–time graph gives the

This parachutist has jumped out of an aircraft at a great enough height for him to enjoy the experience for a period of time. However, he is not in free fall – he has reached a speed where the upward drag of the air on him is equal to the downward pull of gravity, so he goes on falling at this speed.

Key summary: graph of a realistic motion

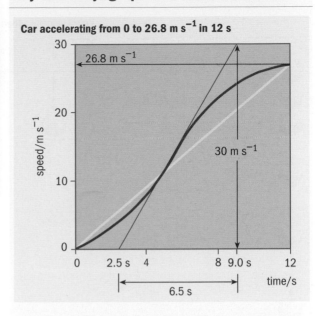

Acceleration is the gradient of the speed–time graph

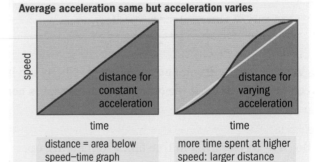

Distance is the area below the speed–time graph

distance travelled.

Graphical methods are useful if you just have observations of distance and time. The speed at any moment can be found by drawing a straight line that has the same slope as the curved distance–time graph at that moment. This is called the tangent to the curve and needs a good eye to get an accurate result. The speed v at this time is worked out by calculating the slope $\frac{\Delta s}{\Delta t}$ of the tangent (chapter 8, p173).

Free fall

Imagine making a sky-dive. First, you have to pluck up the courage to jump out of the aircraft. You fall faster and faster, accelerating downwards. Soon your downward speed is so great, over 100 mph, that you feel a gale of wind seeming to rush upwards past you. The wind tugs at your clothes. Soon its upward drag becomes as big as your whole weight, at which time you no longer fall faster and faster. Your weight and the upward drag balance out and you fall at a steady high speed – 120 mph, or a bit more than $50 \, \text{m s}^{-1}$, is typical. To land safely, you will have to open your parachute. With it open, you fall at a much slower steady speed.

If air wasn't there, parachuting would be impossible. Falling objects would just go on falling faster and faster. You may have seen a video of an American astronaut dropping a hammer and a feather on the Moon's surface. Both reached the ground at the same time. There is no atmosphere

on the Moon – so no air resistance – and both hammer and feather were able to fall freely. As they did so they went faster and faster. They accelerated at a rate called the acceleration of free fall. This is sometimes called the acceleration due to gravity.

On Earth the effect of air resistance on a small dense object can be neglected when it falls through small distances. It gets close to being in free fall, accelerating at close to the local maximum rate, the acceleration of free fall (g). On Earth, g is close to $9.8 \, \text{m s}^{-2}$. On the Moon, g is $1.6 \, \text{m s}^{-2}$.

Try dropping a coin. Watch it carefully. Can you see it accelerating? You probably can't, because the time of fall is very short, and you don't have time to register the different speeds.

Key summary: falling and going sideways

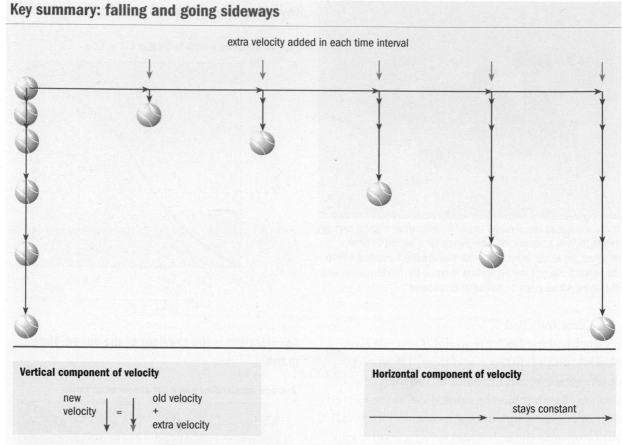

extra velocity added in each time interval

Vertical component of velocity

new velocity = old velocity + extra velocity

Horizontal component of velocity

stays constant

Vertical and horizontal components of velocity are independent; the vertical component increases uniformly with time; the horizontal component is constant

But it is accelerating in free fall. From higher up, say from a bridge overlooking a river, the time of fall is longer. You could drop a small stone and use a stopwatch to time how long it takes to hit the water. Suppose the time of fall is 1.5 s, approximately. The stone falls downwards from zero speed and has an acceleration of $9.8 \, \mathrm{m \, s^{-2}}$. To find the distance, use the relationship:

$$s = ut + \frac{1}{2} a t^2$$

So, $u = 0$, $t = 1.5 \, \mathrm{s}$ and $a = g = 9.8 \, \mathrm{m \, s^{-2}}$. s is the distance down to the water. Thus:

$$s = \frac{1}{2} \times 9.8 \times 1.5^2 = 11 \, \mathrm{m}$$

So you know roughly how high the bridge is.

Of course you can't measure the time to as little as 0.1 of a second, so you know only that the water is about 10 m below the bridge. Even a drop of that

height takes between only one and two seconds.

Ball games, projectiles and parabolas

Waiting to receive a serve at tennis, you know you will have little time to react. As soon as the ball has been hit, it is falling freely. A serve is hit from a height of around 3 m. If simply let drop from that height, the ball would reach the ground in about 0.8 s. If the server hits the ball horizontally at that height, the ball still takes just 0.8 s to reach the ground. Its horizontal velocity simply carries it towards you as it falls. So you have only 0.8 s or so in which to react.

Actually, the server usually hits the ball at a downward angle, giving it quite a large downward component of velocity. So the ball reaches the ground near you even sooner, just as it would if it had been thrown downwards. Air resistance and spin of the ball make some further differences, but the fact remains that tennis is a fast-reaction game.

A tennis player intercepts a ball travelling along a parabolic path.

In these fountains the water streams move in parabolic paths.

The curve followed by the ball is a mathematical parabola, but a tennis ball is no mathematician. It goes along the curve because its free-fall acceleration speeds it up in the downward direction, while its horizontal velocity carries it sideways.

The distance the ball goes horizontally is proportional to the time t. The distance it goes vertically is proportional to t^2. So the shape of the path is the same as that of a graph of t^2 against t, which is a parabola. Or, rather, it has this shape if the forces that are the result of the ball moving through and spinning in the air are unimportant. So a parabola is a good approximation of the path of a moving cricket ball, a fair approximation for a moving football or tennis ball and a very bad approximation for a moving badminton shuttlecock.

If you lob the tennis ball up in the air, your opponent has longer to react, but of course may have to run to get into place to meet the ball coming down again. Here's a way to use vectors to think about the path the ball takes. You send the ball up at an angle. If it weren't for the downward acceleration of gravity, it would go on and on upwards at that angle. But in a short interval of time, the downward acceleration will have added a vertical downward component to the velocity. The resultant velocity is the vector sum of the two. So the ball's motion is tilted down a little, and it slows down a little. In the next moment, the same thing happens, and again, and again, so that the ball's path always curves downwards. Sooner or later, the downward acceleration will have removed all of the upward component of velocity, and the ball will begin travelling downwards.

The ball keeps on doing two things:
- it travels horizontally with the same horizontal component of velocity;
- it accelerates downwards (yes, while going upwards too) with the same acceleration.

That is:
- vectors perpendicular to each other don't affect one another;
- only the vertical component of movement is affected by the acceleration of free fall.

Tracing out such a path is easy to do with a computer drawing package, just copying arrows and putting them together tip to tail. Draw an arrow to represent the initial displacement at a certain velocity, in a short time interval. Continue the same arrow on from the tip of the first to see where the ball would go if there were no gravity. But there is gravity. So add a fixed downward displacement, representing the effect of gravity, to the tip of the second arrow. Now draw in the resultant of these two. This is the predicted displacement in the next time interval. Repeat the process, starting each time from the present displacement. Step by step, a parabola appears.

This is one way in which accelerated motion can be modelled by a computer. The essential point is breaking down the motion into short straight-line steps. The method is called using finite differences. Time is cut into short intervals and in each interval the velocity is imagined as effectively constant. But going from one interval to the next, the velocity is changed. The change in velocity is calculated from the acceleration at that moment. Although used here for constant acceleration, you can probably see

Key summary: lobbing a ball in the air

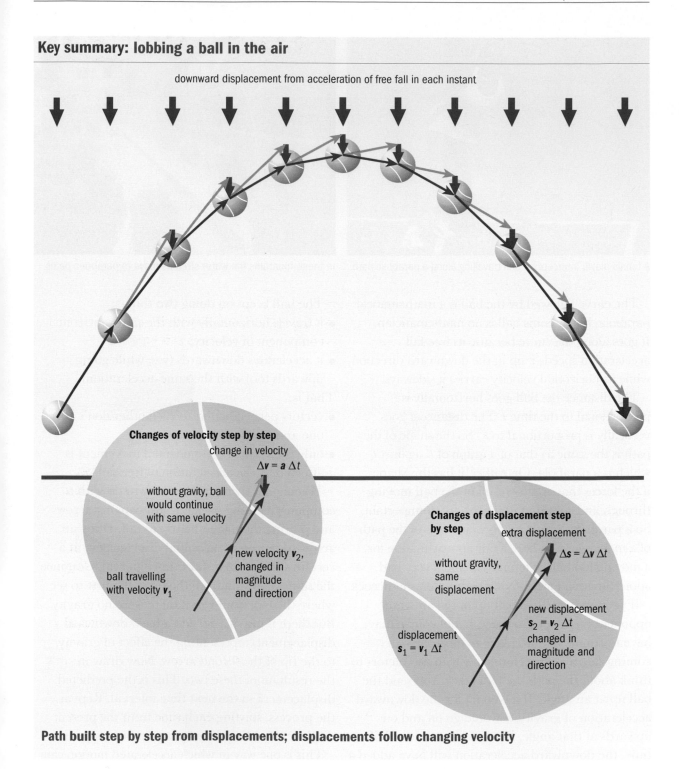

downward displacement from acceleration of free fall in each instant

Changes of velocity step by step

change in velocity

$$\Delta v = a \, \Delta t$$

without gravity, ball would continue with same velocity

new velocity v_2, changed in magnitude and direction

ball travelling with velocity v_1

Changes of displacement step by step

extra displacement

$$\Delta s = \Delta v \, \Delta t$$

without gravity, same displacement

new displacement $s_2 = v_2 \, \Delta t$ changed in magnitude and direction

displacement $s_1 = v_1 \, \Delta t$

Path built step by step from displacements; displacements follow changing velocity

that the method still works if the acceleration varies, as long as you know what it is at each moment.

You may think this step-by-step computation is a bit peculiar. Strictly speaking, it works by imagining switching gravity on and off, which can't be done. As a result, the path computed is kinked, not smooth. In reality, the velocity is changing smoothly all the time, and the path is

a smooth curve. By making the time intervals smaller and smaller, you can get closer and closer to the true curve. That isn't quite all. If the velocity is changing all the time, it isn't quite obvious just which change of velocity to calculate and at what moment to add it in. Computer calculations of paths can go quite badly wrong, for just this kind of reason.

Quick check

1. In a sprint start on a level track a cyclist reaches a speed of $12\,\text{m s}^{-1}$ in 3 s. Show that the average acceleration is $4\,\text{m s}^{-2}$ and that the distance travelled while accelerating is 18 m.

2. A light aircraft has a take-off speed of $60\,\text{m s}^{-1}$. Its engine can produce an average acceleration of $4\,\text{m s}^{-2}$. Show that it takes the aircraft 15 s to reach take off speed and that the minimum length for the runway is 450 m.

3. Write a few sentences discussing what difference it makes if the aircraft in question 2 takes off into a head wind blowing at a speed of $10\,\text{m s}^{-1}$. What about a tail wind of the same speed?

4. Show that a ball must free fall from a height of 4.9 m to take 1 s to reach the ground. A student calculates that the height for a drop taking 100 s is about 50 km. Criticise the calculation.

5. A tennis ball is hit horizontally from a height from which it takes 0.6 s to fall to the ground. Show that
   (a) it was hit at $16.7\,\text{m s}^{-1}$ if it lands 10 m away;
   (b) half way through its flight time it was 1.3 m above the ground.

6. A diver dives off a 10 m high diving board. Show that he reaches the surface of the swimming pool at $14\,\text{m s}^{-1}$ and that this is equivalent to 50 km per hour.

Links to the *Advancing Physics* CD-ROM

Practise with these questions:
40M Multiple choice *Acceleration–time graphs*
50S Short answer *Calculating accelerated steps*
50D Data handling *Analysing motion sensor data*
60S Short answer *Uniform acceleration*
70S Short answer *Braking distance and the Highway Code*
80S Short answer *Throwing a ball*

Try out these activities:
80S Software-based *Constant acceleration with graphs*
90S Software-based *Investigating acceleration with velocity–time graphs*
150S Software-based *Modelling parabolic motion*

Look up these key terms in the A–Z:
Acceleration; distance–time graph; free fall; projectile; speed–time graph

Go further for interest by looking at:
130L Launchable file *Missile and target collide*

Revise using the revision checklist and:
500 OHT *Logic of motion 1*
600 OHT *Logic of motion 2*
700 OHT *Graphs for constant acceleration*
800 OHT *Graphs for realistic motion*
1700 OHT *A parabola from steps*

9.3 Force, mass, gravitation

Sitting at the end of the runway, your aircraft's engines go to full throttle. The brakes are released and the back of your seat pushes you forward as the plane gathers speed for take-off. The holiday has begun. A typical acceleration is $3\,\mathrm{m\,s^{-2}}$, similar to that of a sports car (0–60 mph in 8.9 s), but the aeroplane's acceleration goes on a good bit longer.

Runway lengths at civil airports don't differ much, and take-off speeds of different civil jet aircraft are much the same as one another, so the big planes need to accelerate as fast as the small ones. For that reason, the more massive planes need more powerful engines, which provide more thrust. For example the huge Boeing 747, with maximum take-off mass 375 000 kg, has engines that provide a thrust of more than 1000 kN (the weight of ten 10 tonne lorries). The mass of the much smaller Boeing 737 is about one-fifth of that of a 747, and its engines, in fact, produce about one-fifth of the thrust.

You have met the same thing very often. You can't easily start running quickly carrying a full suitcase, and it's harder still if you are carrying two – both have to be accelerated. You get used to pushing a younger brother or sister on a swing; then one day when they have grown bigger, it's noticeably harder to do. Kick a big stone or hit a big punch-bag, and you hurt your toe or fist, while the stone and the punch-bag hardly move at all. Trains are massive, and despite powerful locomotives they accelerate slowly out of a station.

Putting this all together means that you need a bigger force to give the same acceleration to a bigger mass, and you need a bigger force to give a bigger acceleration to the same mass. Force F, mass m and acceleration a are linked by the simple relationship:

$$\text{force} - \text{mass} \times \text{acceleration or } F = ma$$

Force and acceleration are vectors. Mass is a scalar, not a vector. The usual units are: acceleration a in metres per second per second ($\mathrm{m\,s^{-2}}$); mass m in kilograms (kg); and force F in newtons (N).

The equation above explains why the design thrust of an aeroplane is roughly proportional to its mass. Take-off accelerations are similar, so $F \propto m$.

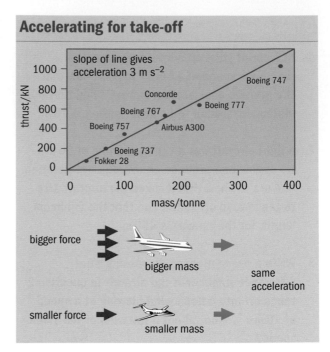

Accelerating for take-off

slope of line gives acceleration 3 m s⁻²

bigger force
bigger mass

smaller force
smaller mass

same acceleration

A sprint start

The relationship force = mass × acceleration helps us to analyse how well an athlete performs. The graph in "Sprint start" (opposite) shows data for a sprinter accelerating from starting blocks, taken from a video. For the first second the athlete's horizontal velocity increases fairly uniformly, increasing by nearly $1.5\,\mathrm{m\,s^{-1}}$. His acceleration is about $1.5\,\mathrm{m\,s^{-2}}$. This sprinter looks fairly big: his mass, at a guess, might be 80 kg. To produce this acceleration would need his leg muscles to produce an average force of 120 N. This isn't all that big a force. It's what you need to pick up a dozen 1 kg bags of sugar.

The sprinter's body weight is 800 N, so the driving force of 120 N is 15% of the force he would need to lift himself, say going upstairs. To accelerate as fast as possible, the sprinter uses starting blocks, which help the initial driving force from his legs to be as much as possible in a horizontal direction, and he leans forwards for the first few paces of the race.

Initial acceleration is the key to winning. The best accelerator will be ahead of the field after the first second or so. The world's best athletes will aim to run 100 m in less than 10 seconds. This is an average speed of at least $10\,\mathrm{m\,s^{-1}}$, so the sprinter must accelerate to a bit more than that speed inside a second or so, and keep it up for the next 9 s.

Clearly the secret to reaching Olympic standard in the 100 metres is a good power:weight ratio. The ideal is a light body with powerful leg muscles. Powerful muscles are big muscles. But big muscles are heavy. Top-class male sprinters are usually large and tall. Longer legs mean more efficient use of muscles (using long bones as levers) and less energy loss because the feet don't hit the ground so often. Each stride also covers more ground.

Let's return to the sprinter in the photograph. The end of the graph suggests that the sprinter has reached a maximum speed of about $2.5 \, \mathrm{m \, s^{-1}}$, so perhaps he wasn't trying very hard.

Mass and weight: on Earth, on the Moon

Two quite different things can happen to a lump of matter, both of which depend on its mass:

- it can be pushed by a force and accelerate – it may be hard to get moving;
- it can be pulled down towards the Earth by gravity – it feels heavy.

Try this: imagine getting ready for an aerobics or gymnastics class. Stand still and just let yourself feel your weight in the soles of your feet. Tell yourself that this feeling is the Earth pulling you down. Try to feel the sensation in your feet, not as you pushing down but as a pull coming from the Earth below. That's the action of gravity: your weight. Now spring a little from side to side. Feel the pushes and pulls you have to give to set your body moving one way, to stop it and start it going the other way. That's due to your mass, needing a force to change its velocity, to accelerate it. Sometimes you'll see this called "inertial" mass (from a Latin word meaning lazy).

Let us tell you what this would feel like if you did it on the Moon, where the gravitational pull is one-sixth of that of the Earth. The feeling of weight on your feet would be one-sixth of what you experienced on Earth – you would need much less effort to lift yourself on your toes. But moving your body to and fro would feel exactly the same as on Earth! So you could easily lift a fellow astronaut off the ground, but it would be just as hard as usual to pull them quickly towards you. In hurdle race on the Moon, you could jump higher but you could not do a much faster sprint start.

If you often went to the Moon, you would have

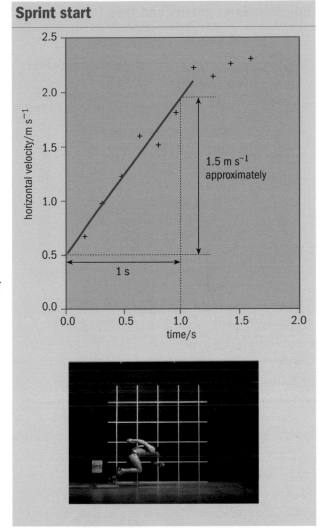

Sprint start

horizontal velocity/m s⁻¹ vs *time/s*

$1.5 \, \mathrm{m \, s^{-1}}$ approximately

1 s

This graph was produced from a video recording of a sprint start. It shows the movement of a point at a runner's waist.

a clear idea of the difference between mass and weight. As it is, staying on Earth, the two seem always to go together and are hard to tell apart. And in shops, the 3 kg mass of a bag of potatoes is always actually called its weight. There's are good reasons why this is what people do. The reasons are that:

- the pull of gravity mg on an object is proportional to its mass;
- the force ma needed to accelerate an object is proportional to its mass.

The result is that the weight is always a good indication of the mass, if the gravity field you live in stays the same. One important difference is that weight mg is a force – the pull of gravity – and has units newtons (N) while by contrast mass m has units kilograms (kg).

Key summary: gravity and free fall

Two things that happen to masses

forces accelerate them
force = mass × acceleration

gravity fields exert a force on them
gravitational force
= mass × gravitational field

if these two masses mean the same

gravitational force
= mass × gravitational
acceleration

acceleration of free fall g
$= \dfrac{\text{gravitational force}}{\text{mass}}$

gravitational field
$= \dfrac{\text{gravitational force}}{\text{mass}}$

acceleration of free fall g = gravitational field

units m s^{-2}

units N kg^{-1}

Units
of acceleration: m s^{-2}
of force: N = kg m s^{-2}

Units
of field: N kg^{-1}
= kg m s^{-2} kg^{-1}
= m s^{-2}

The gravitational field strength can be measured by the acceleration of free fall

The gravity field

The relationship $F = ma$ helps us to understand and predict the motion of cars, aeroplanes, trains and sprinters. But it is interesting to think that the start of it all was with Galileo and Newton. What Newton was trying to understand was movements in the heavens, not getting about on Earth. To explain them, he developed the idea of force being proportional to mass and acceleration. To account for the forces on planets and stars he imagined a force acting in empty space: the force of gravity. The idea is used today to calculate how to place a TV satellite so that millions of satellite dishes can point to it (chapter 11).

Try to picture gravity. You know that everywhere there is a downward pull towards the Earth – it's what makes cups fall and break, and doing the high jump difficult. How do you picture an invisible something everywhere that pulls things downwards? One way is to imagine a downwards arrow at each point in space. You can't draw arrows everywhere, so why not draw them equally spaced? Then these equally spaced arrows of the same length suggest that the pull is the same everywhere throughout space.

This picture is an example of what physicists call a **field**. It is one of the great imaginative leaps made in physics. It led in the end to ideas about electric and magnetic fields, about electromagnetic waves, and to modern theories of the nature of fundamental particles. Yet it seems crazy. How could empty nothingness do anything? No wonder Newton's contemporaries, especially French thinkers, were very critical of it.

Objects of different mass fall freely with the same acceleration. To give a 10 kg mass the same acceleration as a 1 kg mass needs 10 times the force. By coincidence, the force of gravity on the 10 kg mass is exactly 10 times as great, so that the end result is the same: both masses accelerate downwards at the same rate of 9.8 m s^{-2}. A neat coincidence? Not really. It says that mass as measured by accelerations is exactly the same as mass as measured by gravitational pulls.

This means that the strength of the gravity field can be measured by the acceleration of free fall. At least, it can if the effect of the rotation of the Earth can be ignored. The units of gravity field strength are newtons per kilogram (N kg^{-1}). The units of acceleration are metres per second squared (m s^{-2}). They look different, but are really the same.

Forces change velocities

Newton explained that whenever a velocity is changed a net force must be involved. Sometimes the change is simply a change in speed – the moving object gets faster or it slows down. But a velocity can change without a change in speed. This is where the idea that velocity is a vector really bites. Adding a change to a vector can change its direction as well as changing its

Key summary: different masses falling

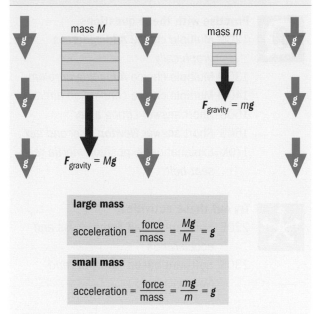

Different masses fall with the same acceleration

large mass

$$\text{acceleration} = \frac{\text{force}}{\text{mass}} = \frac{Mg}{M} = g$$

small mass

$$\text{acceleration} = \frac{\text{force}}{\text{mass}} = \frac{mg}{m} = g$$

Key summary: change of speed and direction

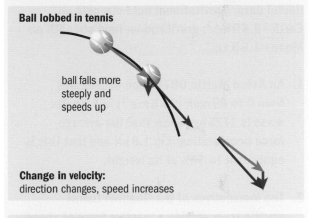

Ball lobbed in tennis

ball falls more steeply and speeds up

Change in velocity: direction changes, speed increases

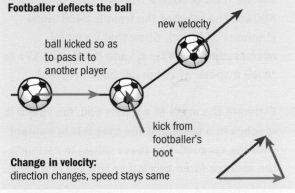

Footballer deflects the ball

new velocity

ball kicked so as to pass it to another player

kick from footballer's boot

Change in velocity: direction changes, speed stays same

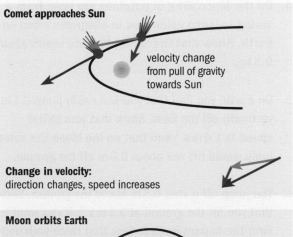

Comet approaches Sun

velocity change from pull of gravity towards Sun

Change in velocity: direction changes, speed increases

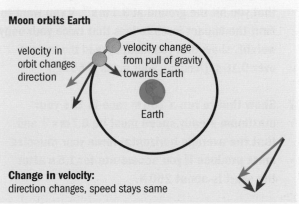

Moon orbits Earth

velocity in orbit changes direction

velocity change from pull of gravity towards Earth

Earth

Change in velocity: direction changes, speed stays same

Changes in velocity change speed, direction or both

magnitude: that's vector addition by the tip-to-tail rule.

So, according to Newton, forces change velocities – sometimes in magnitude (speed), sometimes in direction, sometimes both.

This is connected with an idea of Galileo's that most people find shocking – no force, no change in velocity. But no change in velocity doesn't mean no velocity. If no force acts then things can just keep moving. So the idea is that movement itself needs no force. Just moving isn't special, if only because velocities are all relative.

This is a far cry from the world of horse-drawn carts and coaches being pulled with great difficulty along the muddy and rutted roads of 17th-century Europe. It's an equally far cry from the high-speed trains and aeroplanes of today. You never get the chance to experience the complete absence of forces, so it is hard to believe. Your belief might increase for a moment if you found yourself gliding on skates towards the barrier round an ice-rink, and hadn't learned how to dig in the skates to slow down. Maybe really to see the point you have to look to the stars, like Newton. What keeps them moving? Nothing. They just keep going unless a force changes their velocity.

Quick check

Useful data: Gravitational field strength on Earth $= 9.8\,N\,kg^{-1}$; gravitational field strength on Moon $= 1.6\,N\,kg^{-1}$.

1. An Aston Martin DB7 coupé accelerates from 0 to 60 mph ($26.8\,m\,s^{-1}$) in 5.8 s. Its mass is 1725 kg. Show that the average force accelerating it is 1.8 kN and that this is equivalent to 46% of its weight.

2. The locomotives of the Channel tunnel shuttle train provide a tractive force of about 800 kN. The mass of the train is 2400 tonne (1 tonne = 1000 kg). Show that its initial acceleration is $0.33\,m\,s^{-1}$ and it will take 17 s to reach a speed of 20 km per hour from rest.

3. Estimate the mass of a tennis ball, the speed it reaches in a serve and the time it is in contact with the racquet. Use these values to estimate the average force exerted on the ball in a serve.

4. On the Moon, 5 kg of potatoes are hung from a spring balance calibrated in kilograms when on Earth. Show that the spring balance reads about 0.8 kg.

5. On Earth, you find that you can easily jump 0.1 m vertically off the floor. Show that your initial speed is $1.4\,m\,s^{-1}$ and that on the Moon the same jump would lift you about 0.6 m off the ground.

6. You jump off a step 0.5 m above the ground. Show that you hit the ground at $3.1\,m\,s^{-1}$. If you want to limit the impact force to less that twice your body weight, show that you must spread the impact over 0.16 s. (Assume your mass is 55 kg.)

7. Show that to run a 100 m race in 15 s your maximum steady speed must be $6.7\,m\,s^{-1}$ and that the average horizontal force your muscles must produce if you accelerate for 1.5 s after the start is about 250 N.

Links to the *Advancing Physics* CD-ROM

Practise with these questions:
120M Multiple choice *Adding forces graphically*
130M Multiple choice *A loading problem*
140M Multiple choice *Landing an aircraft*
150S Short answer *Lifting a car*
160S Short answer *Newton's second law*
170X Explanation–exposition *Inertia reel seat belt*

Try out these activities:
220S Software-based *Force, mass and acceleration*
230S Software-based *Transport and Newton's laws*

Look up these key terms in the A–Z:
Free fall; gravitational field; mass; Newton's laws of motion; weight

Go further for interest by looking at:
10T   Text to read *The mysteries of mass*

Revise using the student's checklist and:
900   OHT *Identifying changes in velocity*

9.4 Transport engineering

You are driving down a quiet street. Suddenly a child runs into the road ahead of you. Can you stop in time? Car manufacturers have to design brakes so that a car can stop with better than a minimum deceleration. Using $F = ma$, this means that the greater the mass of the car, the greater the braking force must be.

The Highway Code suggests typical stopping distances for cars. The stopping distance is made of two parts. One part is the distance the car goes while the driver is reacting, before the brakes go on.

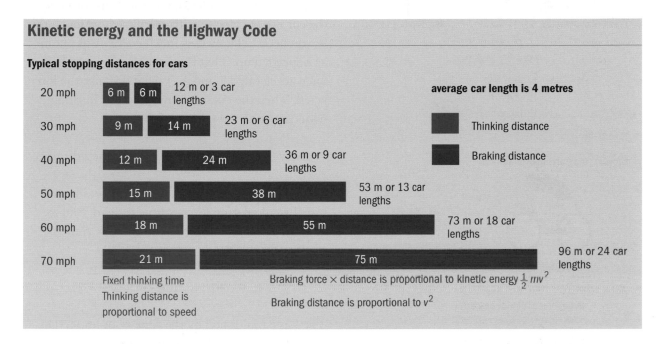

Kinetic energy and the Highway Code

Typical stopping distances for cars

20 mph	6 m	6 m	12 m or 3 car lengths
30 mph	9 m	14 m	23 m or 6 car lengths
40 mph	12 m	24 m	36 m or 9 car lengths
50 mph	15 m	38 m	53 m or 13 car lengths
60 mph	18 m	55 m	73 m or 18 car lengths
70 mph	21 m	75 m	96 m or 24 car lengths

average car length is 4 metres

Thinking distance

Braking distance

Fixed thinking time
Thinking distance is proportional to speed

Braking force × distance is proportional to kinetic energy $\frac{1}{2}mv^2$

Braking distance is proportional to v^2

Key summary: calculating kinetic energy and momentum

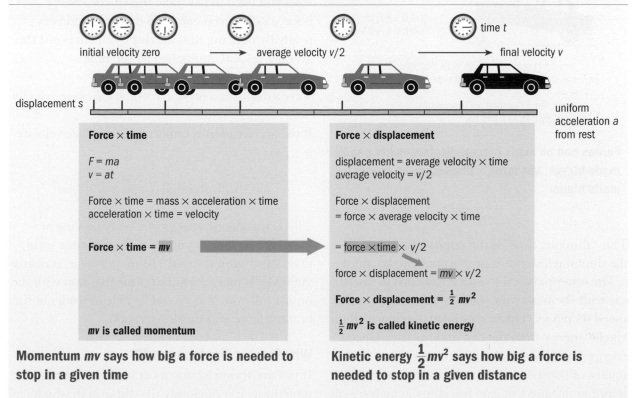

initial velocity zero

average velocity $v/2$

final velocity v

time t

displacement s

uniform acceleration a from rest

Force × time

$F = ma$
$v = at$

Force × time = mass × acceleration × time
acceleration × time = velocity

Force × time = mv

mv is called momentum

Force × displacement

displacement = average velocity × time
average velocity = $v/2$

Force × displacement
= force × average velocity × time

= force × time × $v/2$

force × displacement = $mv \times v/2$

Force × displacement = $\frac{1}{2}mv^2$

$\frac{1}{2}mv^2$ is called kinetic energy

Momentum mv says how big a force is needed to stop in a given time

Kinetic energy $\frac{1}{2}mv^2$ says how big a force is needed to stop in a given distance

Key summary: forces and displacements

Lever: magnifying a force

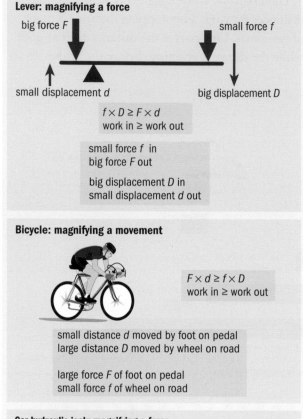

big force F small force f

small displacement d big displacement D

$$f \times D \geq F \times d$$
work in ≥ work out

small force f in
big force F out

big displacement D in
small displacement d out

Bicycle: magnifying a movement

$$F \times d \geq f \times D$$
work in ≥ work out

small distance d moved by foot on pedal
large distance D moved by wheel on road

large force F of foot on pedal
small force f of wheel on road

Car hydraulic jack: magnifying a force

$$f \times D \geq F \times d$$
work in ≥ work out

jack lever moved through large distance D
car lifted through small distance d

small force f exerted on jack handle
very large force F lifts car

Forces can be made bigger; displacements can be made bigger; but force × displacement cannot be made bigger

This "thinking time" is the same at any speed, so the thinking distance is proportional to the speed.

The other part is the distance needed to stop the car with the brakes on. This increases a lot as the speed increases. The brakes have to remove all the kinetic energy from the car, and the car's kinetic energy is proportional not to the speed but to the square of the speed. This explains the very large stopping distances of cars travelling at high speeds.

You can see this from the kinematic equation:

$$u^2 = -2as$$

(with final velocity v equal to zero).

For fixed deceleration, the stopping distance is proportional to u^2. The kinetic energy is the energy the car has at velocity v. To calculate how much this is, you need to know how much energy was put in as it speeded up. That is, you need to know what force accelerated it and over what distance the force acted. Then the energy going from fuel burning with oxygen to the motion of the car is:

energy transferred = work done = force × distance

To calculate force × displacement it is easiest to start by calculating force × time instead. Since $F = ma$, and at is the increase in velocity v, then starting from rest:

force × time = mv

The force multiplied by the time is an important quantity, called the momentum. It tells you how much time you have to stop the car. But what you need to calculate the kinetic energy is force × displacement. This can be found very neatly by noticing that the displacement is just the average velocity multiplied by the time:

force × displacement = force × time × average velocity

If the acceleration is uniform, the average velocity is $\frac{v}{2}$, so:

$$\text{force} \times \text{displacement} = mv \times \frac{v}{2} = \frac{1}{2}mv^2$$

This is the kinetic energy of a body moving at speed v. The kinetic energy of the Eurostar train, mass 2400 tonnes, speed 140 km per hour, is nearly 2000 MJ. It takes 1.5 km to come to a stop with the brakes fully on. You should be able to work out the braking force and the deceleration.

Why force × displacement?

If you are trying to stop a car before you hit something, it is obviously the distance in which

Key summary: going uphill and increasing potential energy

Energy used to go uphill

mass of train m
weight of train mg
vertical distance risen uphill h

mass m

height h

gravitational potential energy increases

Energy to lift the train = force × displacement

increase in gravitational potential energy = **mgh**

force = weight mg
displacement = h

Increase in gravitational potential energy = mgh

you have to stop that matters. But there are more basic reasons why force×displacement is used to calculate work done.

Do you know how to lift someone with just one finger? Stand them on a plank, with a fulcrum under it near their feet, and push down with a finger on the other end. You don't have to push hard, so you don't hurt your finger. But there's a price to pay for using the plank as a lever to magnify the force. While you push down quite a long way, the person is lifted by only a small amount. Forgetting any friction, and the weight of the plank, the force×displacement is the same at both ends. The work you put in comes out unchanged at the other end.

Your bicycle works the other way round. The chain and wheels magnify the distance your feet pedal into a bigger distance travelled by the bicycle wheels. The price is that you have to push harder on the pedals than the force pushing the bicycle forwards. That's why you can't accelerate very fast on a bike. The energy – force×distance – that you put in cannot be increased.

Designing railway gradients

When the railway system was built in the 19th century, large numbers of cuttings, tunnels through hills and embankments above valleys had

to be built. This labour was done with picks and shovels and was expensive, so transport engineers like Isambard Kingdom Brunel planned the routes of their railways carefully, following the contours of the land as much as they could.

The reason to take all this trouble is that trains are very bad at going uphill. Trains are heavy, and it takes a lot of energy to drag them uphill. The amount of energy needed is easily calculated, from force×displacement. If the mass of the train is m, its weight is mg, and if it goes a vertical distance h as it goes uphill, then:

$$\text{force to lift train} = mg$$

$$\text{displacement in direction of force (vertical)} = h$$

$$\text{work done} = \text{force} \times \text{displacement} = mgh$$

$$\text{increase in gravitational potential energy} = mgh$$

The design of the Channel tunnel link is no exception. The tunnel has to dive under the Channel and come back up again, as well as getting through or round hills on the way. The maximum permitted gradient for Eurostar is 1 in 90, so if the track goes up or down by 100 m that height change must be spread over 9 km. The working

Key summary: kinetic and potential energy

mass m

initial velocity = 0

acceleration g
force of gravity mg

height h

velocity v

Falls with constant acceleration:
average velocity = $\frac{1}{2}v$
$v = gt$, so time $t = v/g$
distance = average velocity × time
$h = \frac{1}{2}v \times (v/g)$
$gh = \frac{1}{2}v^2$

force × displacement = mgh

$gh = \frac{1}{2}v^2$
$mgh = \frac{1}{2}mv^2$

Potential energy

mgh

work = force × distance = energy transfer

Kinetic energy

$\frac{1}{2}mv^2$

Energy comes from the gravitational field: decrease in potential energy = mgh; energy now carried by motion of ball: increase in kinetic energy = $\frac{1}{2}mv^2$; decrease in potential energy = increase in kinetic energy

mass of Eurostar is 2400 tonnes, a weight mg of about 24 MN. Its locomotives deliver a maximum of 11.2 MW, or 11.2 MJ per second. Suppose about half its power, say 6 MJ per second, is being used to go uphill, with the rest used to overcome resistance to motion. Suppose the train is going up a slope and is rising vertically by distance h each second:

$$\text{energy provided per second} = 6\,\text{MJ per second}$$

$$\text{energy needed per second} = 24\,\text{MN} \times h$$

$$h = \frac{6\,\text{MJ per second}}{24\,\text{MN}} = 0.25 \text{ m per second}$$

A gradient of 1 in 90 will spread this 0.25 m rise over a distance of about 22 m. If these guesses are correct, it seems that the train can't go uphill at more than 22 m s^{-1}, or 80 km per hour. Its locomotives can't increase gravitational potential energy any faster.

Energy exchanges

Having brought Eurostar up to speed and given it a lot of kinetic energy, or having taken it uphill and provided a lot of gravitational potential energy, it would be a pity to throw all that energy away when the train slows down or runs downhill. And it isn't thrown away. To slow the train down, its electric motors are switched to act as dynamos, and feed at least a part of its kinetic energy back into the overhead electrical supply lines. So, going downhill Eurostar can provide electrical power back to the national grid, as the gravitational potential energy decreases. The train to Paris may be helping to keep your lights on, even at this moment.

Energy exchanges between gravitational potential energy and kinetic energy are very common. They happen every time a ball falls under gravity. As it falls, gravitational potential energy comes from the gravitational field and goes to increasing the kinetic energy of the ball. If no energy goes to stirring up the air on the way, the gravitational potential energy decreases by mgh, and the kinetic energy increases by $\frac{1}{2}mv^2$ (if the ball started at rest). The changes in both can be calculated by the work done, force × displacement.

You know that energy from burning fuel is used to give kinetic energy to cars and trains. But energy can also come from gravity. Energy is needed to lift something away from Earth and, when it falls, that energy can be retrieved. Gravitational potential energy acts as a store of energy.

Key summary: flow of energy to and from a train

Travelling at constant high speed

energy flows to train

25 000 V overhead wire

air in front of train set in motion

kinetic energy of train stays constant

all the energy flows from train, setting air in motion, overcoming friction at axles and deforming rails

Accelerating

energy flows to train

kinetic energy of train increases

only part of energy flows from train, setting air in motion, overcoming friction at axles and deforming rails

Slowing down

energy returned to supply

kinetic energy of train decreases

energy flows from train, brakes acting, setting air in motion, etc.

Change in kinetic energy = energy in − energy out

A trip to Paris: powering a high speed train

Travelling to Paris one spring weekend, your mind is full of plans for what you are going to do. Inside the train, you'd hardly know you were moving, even though you see the countryside flashing past out of the corner of an eye. It's all so smooth that you don't think about the large flows of energy to and from the train. But think now about what is happening. As the train pushes into the air in front of it, it sets that air moving, probably at a speed similar to that of the train. A surprising amount of air is set moving. The train can travel least 50 m in a second (180 km per hour). Its frontal area might be 20 m^2. That would be 1000 m^3 of air pushed aside per second, with a mass of 1.3 tonnes (1300 kg). Set moving

at 50 m s^{-1} the energy $\frac{1}{2}mv^2$ comes to 1.6 MJ per second, or 1.6 MW.

Add to this the drag of air on the sides of the train, friction heating the axles and rails, and energy used in deforming the track as the heavy train goes over it, you soon reach the 10 MW or so of power needed to keep the train running.

When the train is running at a steady speed, energy flows out as fast as it flows in. If the driver increases the input power, more energy comes in than goes out, and the difference increases the kinetic energy of the train. If the driver turns off the input power and puts on the brakes, energy goes rapidly from the train to the surroundings, and its kinetic energy decreases.

The power needed to keep the train running at

Key summary: power, force and velocity

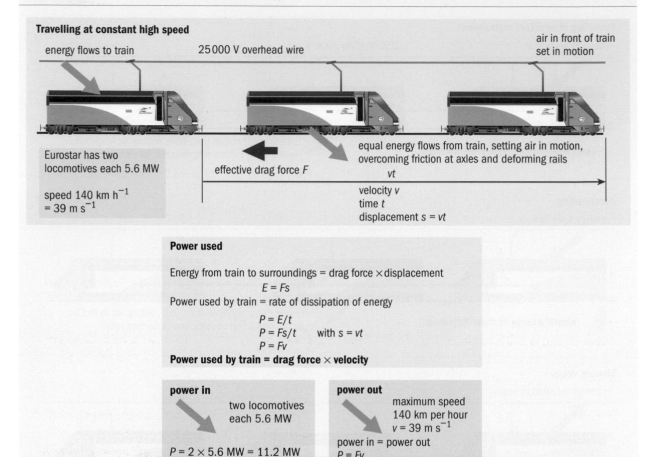

Travelling at constant high speed

energy flows to train   25 000 V overhead wire   air in front of train set in motion

Eurostar has two locomotives each 5.6 MW

speed 140 km h^{-1}
= 39 m s^{-1}

effective drag force F

equal energy flows from train, setting air in motion, overcoming friction at axles and deforming rails
vt

velocity v
time t
displacement $s = vt$

Power used

Energy from train to surroundings = drag force × displacement
$$E = Fs$$
Power used by train = rate of dissipation of energy
$$P = E/t$$
$$P = Fs/t \quad \text{with } s = vt$$
$$P = Fv$$

Power used by train = drag force × velocity

power in

two locomotives each 5.6 MW

$P = 2 \times 5.6 \text{ MW} = 11.2 \text{ MW}$

power out

maximum speed 140 km per hour
$v = 39$ m s^{-1}

power in = power out
$P = Fv$
$11.2 \times 10^6 \text{ W} = F \times 39 \text{ m s}^{-1}$
$F = 290$ kN

Power = rate of transfer of energy = force × velocity

constant speed is just the power needed to replace all the energy the train is losing each second:

work done = drag force × displacement

work done per second = drag force × distance moved per second

But power = work done per second, and velocity = distance moved per second, so:

power = drag force × velocity or $P = Fv$

Imagine a modern Galileo

Galileo argued that things can go on moving forever if no force acts on them. Imagine yourself as a present-day Galileo working as a railway engineer. You know that improving the streamlining of a train reduces drag and so reduces the tractive force needed to keep it going. Imagine super-streamlining, which reduces air drag even more, and super-lubricants, which remove axle friction. Make the rails smoother and smoother and very rigid. Soon you can imagine a train running at hundreds of kilometres per hour off only a car battery.

Then the last step. Imagine that drag is reduced to zero. Then the train will just go on running, with nothing to stop it, and no need for anything to power it. Impossible? Certainly it's only a dream for trains, but look at the stars. Think of the Earth turning day after day. Motion without forces just goes on and on. Motion at constant velocity needs no cause to keep it going.

Try these

1. Show that the kinetic energy of:
   (a) a 100 g tennis ball travelling at $30 \, \text{m s}^{-1}$ is 45 J;
   (b) a 50 kg runner doing a 1000 m race in 4 minutes is 435 J;
   (c) a 0.5 tonne car travelling at $2 \, \text{m s}^{-1}$ is 1 kJ.

2. Show that the change in gravitational potential energy when:
   (a) a 50 kg athlete jumps 1.5 m above the ground is about 750 J;
   (b) a 1 tonne car is jacked 100 mm above the ground is about 1000 J;
   (c) a 200 g ball is hit 20 m up in the air is about 40 J.

3. A car of mass 1235 kg accelerates at $2.1 \, \text{m s}^{-1}$ over a distance of 50 m. Show that the tractive force acting is 2.5 kN and that the final kinetic energy aquired is 125 kJ. Hence show that the final speed is $14.2 \, \text{m s}^{-1}$.

4. A train of mass 1000 tonnes starts going uphill at $10 \, \text{m s}^{-1}$. The extra power required to keep it going is 1 MW. Show that the steepness of the slope is about 1 in 100 (i.e. a 1 m rise for every 100 m horizontal distance).

5. If a swimmer generating 200 W of swimming power can swim at a speed of $0.5 \, \text{m s}^{-1}$, show that the effective resistive drag of the water on their body is 400 N.

6. A tennis player practising follow through in their service manages to keep their racquet in contact with the ball over 0.2 m of their stroke. The ball flies off at $30 \, \text{m s}^{-1}$. If the mass of the ball is 100 g, show that they achieved an average force of 225 N on the ball.

Links to the *Advancing Physics* CD-ROM

Practise with these questions:
240W Warm-up exercise *Along the flat and up the hill*
200E Estimate *A bouncing ball*
210E Estimate *Landing heavily*
220S Short answer *A skateboarder*

Try out these activities:
310S Software-based *Storing energy in springs*
360S Software-based *Energy flows and motion*

Look up these key terms in the A–Z:
Kinetic energy; potential energy; power; work

Go further for interest by looking at:
1500 OHT *Your toolkit for computing the next move*

Revise using the revision checklist and:
1000 OHT *Gravity and free fall*
1100 OHT *Kinetic and potential energy*
1200 OHT *Calculating kinetic energy*
1300 OHT *Flow of energy to a train*
1400 OHT *Power, force, velocity*

Summary check-up

Vectors ✓

- Displacement, velocity, force and acceleration are all vector quantities, adding tip to tail
- The relative velocity of two objects can be found by subtracting the velocity of one of them from both

Kinematics ✓

- Paths of bodies in accelerated motion can be constructed step by step using small finite time intervals
- Displacements and velocities for uniform acceleration can be predicted using the kinematic equations:

$$v = u + at$$

$$s = \frac{u+v}{2}t$$

$$s = ut + \frac{1}{2}at^2$$

$$v^2 = u^2 + 2as$$

- The area under a speed–time graph can give the distance travelled
- The slope of a speed–time graph can give the acceleration
- The slope of a distance–time graph can give the speed
- A projectile in free fall follows a parabolic path
- Vertical and horizontal components of velocity of a projectile are independent

Dynamics ✓

- Force, mass and acceleration are related by the equation $\mathbf{F} = m\mathbf{a}$
- The gravitational force on an object is proportional to its mass
- Different masses have the same acceleration g due to gravity
- The gravitational field g is the gravitational force per unit mass

Energy and power ✓

- The kinetic energy of a mass m moving at velocity v is $\frac{1}{2}mv^2$
- The increase in gravitational potential energy for a mass m lifted through a height h in a uniform gravitational field g is mgh
- Work done measures energy transferred and is given by the net force acting multiplied by the component of the displacement in the direction of the force
- In motion under gravity, with no other forces acting, changes in gravitational potential energy are matched by equal and opposite changes in kinetic energy
- Power expended = rate of doing work = force × velocity (in the direction of the force)

Questions

Useful data: $g = 9.8\,\mathrm{N\,kg^{-1}} = 9.8\,\mathrm{m\,s^{-2}}$.

1. **This question is about free fall under gravity.**
   (a) Round shot for use in old muskets was made by dropping molten iron down the inside of a tall tower. The iron formed a sphere and then solidified before falling into cold water. If it takes 3 s for the iron to solidify how high must the tower be (at least).
   (b) You are standing under a bridge and want to know how high it is above you. You (carefully) throw a small stone up at just the right speed to barely reach the bridge. You repeat the measurement several times and find its time of flight (up and down) to be 2.4 s on average. How high is the bridge?

2. **Galileo did experiments on the motion of objects. In a series of experiments he rolled a metal ball down a sloping board, changing the angle of the board to vary the results. The data from one such experiment (in modern units) are as follows: length of board = 3 m; angle of board to the horizontal = 30°; time taken for ball to roll down board = 1.2 s.**
   (a) What was the average acceleration of the ball?
   (b) g acts vertically downwards. What is the component of g that acts along the board to produce the acceleration?
   (c) From your answers to (a) and (b) what value of g did Galileo obtain? Suggest two reasons why the final result for g is less than the modern value.

3. **The following data relate to a modern car. Time to reach 60 from 0 mph = 9.9 s; kerb weight = 1150 kg; distance needed to brake from 30 to 0 mph = 11.1 m; distance needed to brake from 70 to 0 mph = 63.6 m**
   (a) Convert the speed data to SI units.
   (b) Calculate the acceleration of the car from 0 to 60 mph in $\mathrm{m\,s^{-2}}$.
   (c) What is the average driving force on the car needed to produce this acceleration?
   (d) What value of deceleration was required to brake from 30 to 0 mph in a distance of 11.1 m?
   (e) What braking force is needed to produce this deceleration?
   (f) What value of deceleration was required to brake from 70 to 0 mph in a distance of 63.6 m?
   (g) What braking force is needed to produce this deceleration?
   (h) Use the data to explain to a learner driver why it is dangerous to exceed the speed limit of 30 mph in built-up areas.

4. **The data in table 1 were produced while testing a new type of car.**

Table 1

t/s	0	0.2	0.4	0.6	0.8	1.0	1.2	1.4	1.6	1.8
v/m s^{-1}	0	0.64	1.28	1.92	2.56	3.20	3.84	4.4	5.12	5.76

   (a) Draw a speed–time graph for these data.
   (b) Describe the motion of the car in the first 1.8 s.
   (c) What was the acceleration of the car?
   (d) How far did the car travel in the first second?

5. **In skeet shooting, a clay disc is shot into the air at an angle. In one trial of the system the disc is shot at an angle of 60° to the level ground with a speed of 40 m s^{-1}.**
   (a) Resolve this starting velocity into its vertical and horizontal components.
   (b) Use its upward component of velocity to predict that the disc should stay in the air for about 7 s. Is this likely to be an underestimate or an overestimate? Justify your decision.
   (c) How far should the disc have travelled horizontally in this time?

6. **The two engines of a 35-seater propeller aircraft can each produce 1000 kW of power for take-off. The take-off mass is 13 000 kg. The take-off speed is 60 m s^{-1} and the length of runway needed is 600 m. Assume that the aircraft accelerates uniformly up to take-off.**
   (a) What is its acceleration?
   (b) What time does the take-off occupy?
   (c) What is the thrust of the propellers on the aircraft?
   (d) What is the kinetic energy of the aircraft at take-off?
   (e) Calculate the kinetic energy in a different way, from the same data.
   (f) What is the power provided to increase the kinetic energy of the aircraft during take-off?
   (g) Suggest reasons for the difference between the answer to (f) and the power rating of the engines.

Case studies: quality of measurement

This chapter provides a set of case studies about measurement. They illustrate essential ideas about:

- how to measure quantities as well as possible
- how to know how well you have done
- the problems of making good measurements

The case study on the CD-ROM

Replacing mercury thermometers

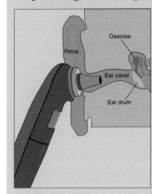

explains the advantages of infrared ear thermometers, which respond in less than a second compared with some minutes for mercury ones, but that are still not officially used by nurses because of a need for calibration

The case studies in the book

The ocean from space

describing how the height of the ocean surface can be measured to within a few centimetres by radar from a satellite, and how this requires corrections for many systematic errors

Calibrating ultrasound transducers

describing a new way to calibrate ultrasound transducers, used in body scanning and in physiotherapy, that is both cheaper and has greater sensitivity than the orthodox method

The most expensive zero error ever

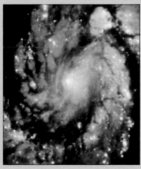

how an accidental zero error of 1.3 mm in a length measurement caused a fault in the Hubble Space Telescope that cost hundreds of millions of dollars to correct

Discovering a natural nuclear reactor

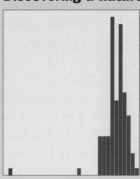

tells the story of the discovery that, some 2 billion years ago, there was a natural nuclear fission reactor in uranium deposits in western Africa

Case study: the ocean from space

Scientists, politicians and the general public are all becoming increasingly concerned about changes to the planet's climate. One factor of particular importance is the behaviour of the world's oceans. International teams of scientists are using satellites to monitor changes to the oceans from space. One result of these observations is that changes in the level of the seas can be measured to within a few centimetres. This requires extremely high resolution, precise measurement of time intervals and careful consideration of systematic errors.

Key ideas: resolution, systematic errors, error correction

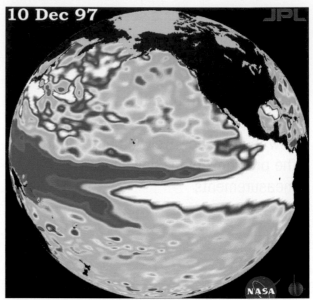

The white areas on this image show where the sea surface is between 14 and 30 cm above the normal level. This indicates a large mass of warm water that moves eastward to the coast of South America and causes an El Niño event.

El Niño

In the last months of 1997 Peruvian anchovy fishermen reported a dramatic decline in fish stocks – El Niño had arrived. Named "the Christ child" in Spanish, the warm current of water arrived along the coast of Peru in time for Christmas. In years when El Niño does not occur there is an upwelling of cooler water near the coast, bringing to the surface water that is rich in the plankton on which anchovies feed. El Niño stops this upwelling and so anchovy stocks plummet.

But the effect of El Niño goes far beyond anchovies. In 1997 the rainfall in Peru was 15 times the average, causing flooding and landslides, and malaria and cholera broke out. The cost of the damage was estimated at $3.6 billion (£1.8 billion). The 1997–8 event also led to droughts in Indonesia and torrential rain and mud slides in California. Millions of people were affected.

It now seems that it may be possible to predict such events. El Niño is linked to a change in the size of the pool of warm water that is found in the Pacific Ocean around the equator. The expansion of the water due to this warming causes the sea

level to rise by a few centimetres. Incredibly, this tiny change in sea level can be detected by a satellite orbiting 1000 km above the ocean.

Sea-level measurement by satellite

This is the story of how a satellite can measure the height of the sea with a precision of a few centimetres. It shows how international co-operation and global systems are needed for these measurements. The principle is simple. Between 1992 and 2006 the satellite system *Topex/Poseidon* took measurements of the sea surface height using radar altimetry. A sensitive altimeter can find distances precisely by measuring the time between the transmission of a pulse of radar (frequency 13 GHz) and the detection of the return pulse reflected by the surface of the water. A simple calculation will give a measurement of the distance between the satellite and the sea.

Time and frequency measurements are as good as measurements get

The resolution needed to measure a 1 cm change in sea level from a distance of 1000 km is one

Key summary: resolution of a radar altimeter

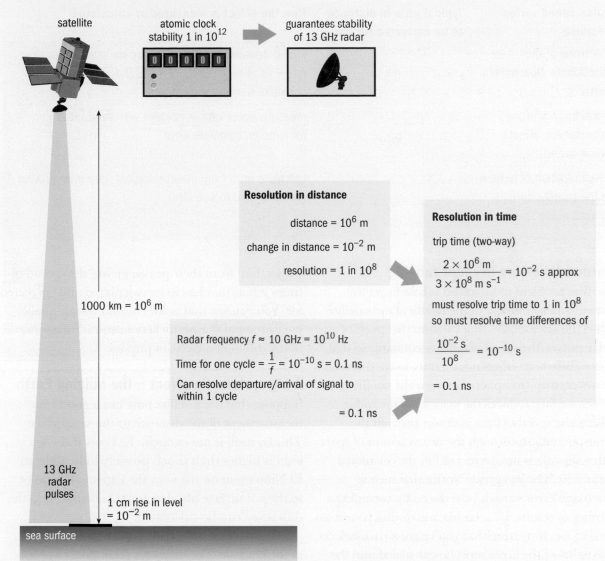

satellite

atomic clock
stability 1 in 10^{12}

guarantees stability
of 13 GHz radar

Resolution in distance

distance = 10^6 m

change in distance = 10^{-2} m

resolution = 1 in 10^8

Resolution in time

trip time (two-way)

$$\frac{2 \times 10^6 \text{ m}}{3 \times 10^8 \text{ m s}^{-1}} = 10^{-2} \text{ s approx}$$

must resolve trip time to 1 in 10^8
so must resolve time differences of

$$\frac{10^{-2} \text{ s}}{10^8} = 10^{-10} \text{ s}$$

= 0.1 ns

1000 km = 10^6 m

Radar frequency $f \approx$ 10 GHz = 10^{10} Hz

Time for one cycle = $\frac{1}{f}$ = 10^{-10} s = 0.1 ns

Can resolve departure/arrival of signal to within 1 cycle

= 0.1 ns

13 GHz
radar
pulses

1 cm rise in level
= 10^{-2} m

sea surface

High resolution plus high stability = high precision (low uncertainty)

part in 100 million. That's a finger's width in the distance across France. This can only be done because measurements of time and frequency are the best kind of measurement that physicists have available to them. This is because of the development of astonishingly stable atomic clocks, which keep time to better than one part in 10^{12} – one second in 30 000 years. The very best ultrastable clocks can today do 10 or 100 times better than this. So, whenever they can, physicists use time or frequency in their measurements of fundamental quantities.

Tracking down systematic error: the price of high resolution

If you make a rough measurement you don't need to worry much about systematic error because one big enough to matter will be pretty obvious, but the higher the precision of your measurement, the more small effects there can be to spoil it. A lot of hard work has to go into thinking of possible systematic errors, measuring or calculating them, and removing them from the results so that any error left is smaller than the uncertainty. Only then will the measurement be accurate as well as precise.

Factors affecting the speed of a radar pulse

Pulse speed varies because	Typical error in distance to be corrected for /m	How the effect is measured or calculated
electrons in the ionosphere slow pulses down	0.5	speed depends on frequency, so measure radar time intervals at two frequencies (13.6 and 5.3 GHz), and calculate density of electrons
water vapour in the troposphere affects pulse speed	0.5	measure water vapour content with on-board radiometer, calculate error
concentration of oxygen in the troposphere affects pulse speed	2	calculate error from meteorological data from ground stations, sent to satellite

Systematic errors in radar ranging

The first problem to be solved is how to get an accurate measurement of the height of the satellite above the sea surface. In a vacuum the speed of radar pulses (the speed of light) is constant, so that a time difference translates directly into a distance. However, the atmosphere between the satellite and the sea surface reduces the velocity of the radar pulse – the speed of light is slower through the atmosphere than through the near-vacuum of space. If this slowing is not corrected for, the calculated distance will be too great. Notice that such a systematic error cannot be reduced by averaging a number of results. It's a bit like measuring your mass as 63 kg but forgetting that you've got a rucksack on. The value of the error must be calculated and the time interval changed accordingly.

Unfortunately this does not boil down to one simple correction for "the speed of the pulses in the air". Their speed is affected by several factors, all important at the high precision required. The table above shows some of those factors, quantifies their effect on measurement and indicates what has to be done to correct for them.

The corrections so far have assumed that the sea surface is flat with no waves. The radar pulse spreads over a region of ocean and the existence of waves in this region makes the return signal spread out in time. Taking the mid-point of the spread-out pulse should cancel out the variations due to wave height. But this doesn't quite work. Radar pulses are reflected more strongly from the troughs of

waves than from their peaks, giving the spread of times a bias that has to be calculated and corrected for. You can see that achieving very-high-quality measurement is a pernickety business, involving many different aspects of physics.

A big systematic effect – the bulging Earth

Suppose that the satellite now has a good measurement of the distance to the sea surface. This by itself is not enough. To know if the sea level is higher than usual, possibly indicating an El Niño event on the way, the expected distance to the sea surface also has to be known so that the difference can be calculated.

The rotation of the Earth makes it bulge out at the equator. The diameter from pole to pole is about 43 km less than the diameter at the equator. This shape is called an ellipsoid. The shape of the average surface of the oceans, called the geoid, does not quite follow this ellipsoid because there are bumps and hollows in the sea surface caused by the presence of undersea mountains and oceanic trenches. Because of this, the satellite carrying the radar altimeter has to know exactly where it is over the surface of Earth and calculate the expected distance to the average sea surface at that place.

Additionally, the height of the sea surface varies from the average because of tides, winds and changes in atmospheric pressure. These changes produce the dynamic topography of the sea surface, so the satellite has to know about tidal conditions and the local weather, and again

The position and coverage of DORIS sensors

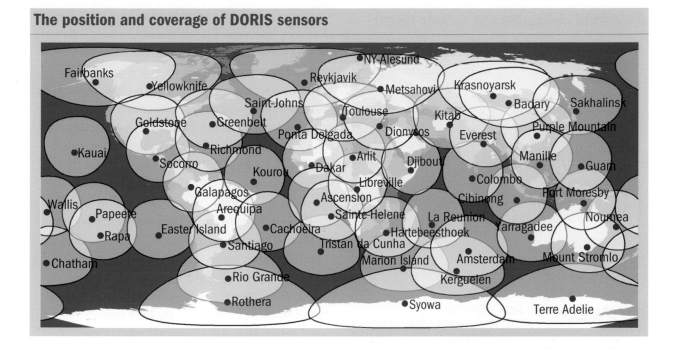

calculate corrections to the expected distance.

For all this to be possible, the satellite has to know very precisely where it is above Earth.

Finding the satellite's position

The satellite's position can be determined to an uncertainty of about 3 cm. This precision is achieved by using three separate methods of orbit tracking. A laser reflector measures the distance from the satellite to ground stations, a global positioning system (GPS) receiver measures the distance to other satellites of known position, and a Doppler orbitography and radiopositioning integrated by satellite (DORIS) receiver measures the motion of the satellite relative to a ground-based station. Of course, all these measurements also need to be corrected for systematic errors.

The satellite's position must be known at all times. This requires more than 50 DORIS sites scattered around the globe.

The satellite carries an on-board navigation system that predicts the position of the satellite, corrects this value with data from DORIS and transmits the corrected position to users. It does this every 10 seconds.

To measure the height of the sea the satellite needs two measurements – the satellite altitude above the reference ellipsoid is found from data from ground-based stations and orbit calculations;

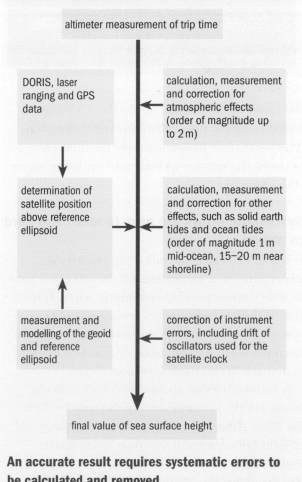

Key summary: error detection and removal

altimeter measurement of trip time

DORIS, laser ranging and GPS data

calculation, measurement and correction for atmospheric effects (order of magnitude up to 2 m)

determination of satellite position above reference ellipsoid

calculation, measurement and correction for other effects, such as solid earth tides and ocean tides (order of magnitude 1 m mid-ocean, 15–20 m near shoreline)

measurement and modelling of the geoid and reference ellipsoid

correction of instrument errors, including drift of oscillators used for the satellite clock

final value of sea surface height

An accurate result requires systematic errors to be calculated and removed

A satellite measuring sea surface height

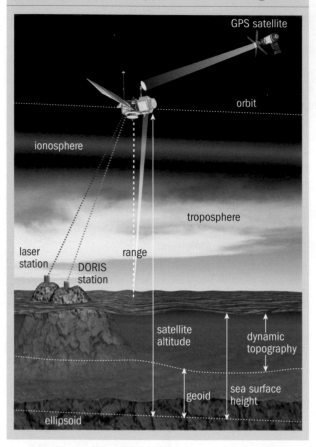

the range of the satellite above the sea surface is found using the principle of radar echolocation. The sea surface height is simply the difference between the satellite altitude and the range above the sea surface.

Systematic errors are easy to make and hard to detect

It is a sad fact that the more you reduce the uncertainty in a measurement, the more you have to worry about possible systematic errors. If the uncertainty is large, only a few sources of large systematic errors matter, but as the measurement becomes more refined, so large numbers of small systematic errors have to be considered.

A cautionary tale about this comes from the history of the measurement of the speed of light. By 1926, Albert Michelson, measuring the speed of light between mountain tops 35 km apart, could claim an uncertainty of only $\pm 4\,\mathrm{km\,s^{-1}}$ in the measured speed of light ($c = 299\,796\,\mathrm{km\,s^{-1}}$). But because of this precision, he had to correct

for the fact that light travels more slowly in air than in a vacuum. The difference to be corrected for was some $67\,\mathrm{km\,s^{-1}}$, a lot larger than his claimed uncertainty. The trouble was that the correction was itself uncertain, because of varying atmospheric conditions. So he tried to avoid the problem by measuring the speed in an evacuated tube and, in 1932, published a new, lower value of $299\,774 \pm 4\,\mathrm{km\,s^{-1}}$, $22\,\mathrm{km\,s^{-1}}$ slower than the mountain-top measurement. This value was internationally adopted until, in the 1940s and 1950s, results from radar measurements suggested that it was too low.

In 1973, after many new measurements, a higher value of $c = 299\,792\,458 \pm 1\,\mathrm{m\,s^{-1}}$ was adopted. Michelson's "better" value had been wrong, but the reason is still not properly understood. One possibility is that the evacuated tube, buried near the seashore, was compressed and expanded as the tide came in and out.

Putting it all together

We have only briefly considered systematic errors due to atmospheric effects. There are many more possible sources of systematic error, all of which must be estimated to remove them from the final result. This is a complex task requiring many people working together to identify, quantify and compensate for such errors. The goal of a better understanding of the planet's oceans is surely worth the considerable international effort.

Summary

- High resolution means that very small changes can be detected and measured
- A radar device with its frequency stabilised by an atomic clock gives highly repeatable measurements
- High resolution and high stability together allow for low uncertainty (high precision)
- Systematic errors have to be measured or calculated and removed if a measurement is to give a true value
- The lower the uncertainty achieved, the more different physical effects there are likely to be leading to a systematic error requiring correction

Case study: calibrating ultrasound transducers

It is important to know the power of ultrasound transducers used in medicine in order to protect patients. The standard way to measure their power is to weigh the force that the waves exert on material that absorbs them. This method of calibration is both insensitive and costly. The National Physical Laboratory is developing a cheaper method where the sensitivity can be controlled and increased.

Key ideas: calibration, sensitivity, noise

Ultrasound in medicine and therapy

One use of ultrasound in hospitals is to monitor the progress of babies in the womb. For this the ultrasound power must not be too great because ultrasound energy can warm tissue that absorbs it. The total power used for scanning a foetus is only a few milliwatts. Even so, care obviously has to be taken not to warm up an unborn baby. This is one reason why ultrasound scanners need to be calibrated and used carefully.

In physiotherapy, ultrasound is used to deliberately warm up possibly damaged muscle tissue to help it heal. Here ultrasound power up to about 10 W is needed. Again, calibration is important from the point of view of safety and to control the treatment given.

With a wide range of powers to measure it is important that the measuring devices used have appropriate sensitivity. They must be sensitive enough to determine small powers but not to be overwhelmed by larger powers.

Weighing ultrasound power

The current standard method for measuring ultrasound power is remarkably simple. You just "weigh" the power. How?

A target slab of a good absorber of ultrasound is placed underwater and weighed with and without the ultrasound power switched on. The difference

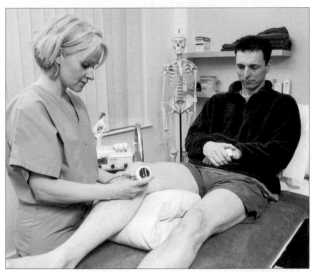

An ultrasound device being used in hospital for physiotherapy.

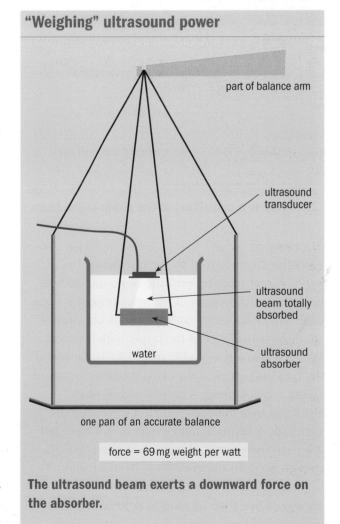

"Weighing" ultrasound power

part of balance arm

ultrasound transducer

ultrasound beam totally absorbed

ultrasound absorber

water

one pan of an accurate balance

force = 69 mg weight per watt

The ultrasound beam exerts a downward force on the absorber.

Key summary: calculating the force

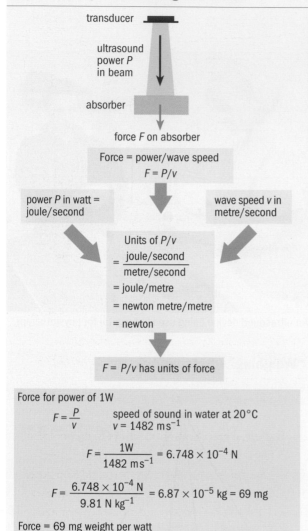

Force for power of 1W

$$F = \frac{P}{v} \qquad \text{speed of sound in water at } 20°C$$
$$v = 1482 \text{ ms}^{-1}$$

$$F = \frac{1W}{1482 \text{ ms}^{-1}} = 6.748 \times 10^{-4} \text{ N}$$

$$F = \frac{6.748 \times 10^{-4} \text{ N}}{9.81 \text{ N kg}^{-1}} = 6.87 \times 10^{-5} \text{ kg} = 69 \text{ mg}$$

Force = 69 mg weight per watt

The force is proportional to the power of the beam

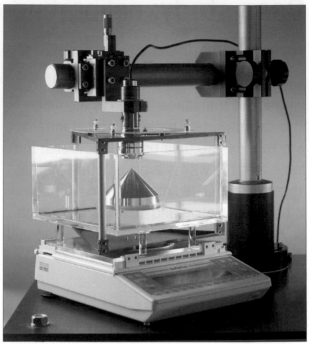

The National Physical Laboratory version of a weighing device for measuring ultrasound power.

gives the force exerted by the ultrasound beam.

The target must reflect little or no ultrasound back to the transducer. The measurement is done underwater because water transmits ultrasound from the transducer to the target with little loss. Water is also a reasonable model for tissue and is the best and simplest standard medium we have.

Any type of energy-carrying wave, when absorbed by a material, exerts a force on that material. A beam of light exerts such a force but it is extremely small. (Its effects are big enough, though, to push material away from a comet to make its tail.) An ultrasound beam exerts a stronger force than a beam of light of the same power. This force comes close to 70 mg weight

per watt. (That is, 1 W of power gives a force equivalent to the weight of about 7×10^{-5} g, which is about 7×10^{-4} N.) Such a force can be measured accurately on a good enough balance. So the sensitivity of this method of measuring ultrasound power is about 70 mg weight per watt. Remember, the sensitivity of any instrument is just the ratio of its output to its input.

Force, power and sensitivity

It is easy to calculate the sensitivity of the weighing method. The relationship between the power of a beam of waves and the force exerted when they are all absorbed could hardly be simpler. It is

$$\text{force} = \frac{\text{power}}{\text{wave speed}}$$

You can see that the units in this equation are correct – the ratio of power to speed does have units of force.

The greater the wave speed, the weaker the force for a given power, hence a light beam exerts only a very tiny force. The speed of ultrasound in water is just less than 1500 ms^{-1} and you can see that the force exerted for a power of 1 W is 69 mg weight. For a 100 mW ultrasound scanner it would be 10 times less – not quite 7 mg weight – so a very sensitive balance is necessary.

Such balances are rather expensive, typically £2000. They also require considerable skill to use. Even for the higher values of power to be checked, say 10–15 W, the force is still only a fraction of a gram in weight. The calibration of ultrasound transducers is likely to be checked less often than it should be because it is expensive and difficult.

The source of the problem is the inherent sensitivity of the method. The small force involved can only be measured accurately with an expensive and bulky balance. Might there be a better way that gives more control over the sensitivity? The National Physical Laboratory (NPL) thought that there might be.

The navy comes to the rescue

NPL is developing a new method for measuring ultrasound power based on the heating effect of the energy absorbed. The idea is to have a target material good enough to absorb all the ultrasound energy within a millimetre or so of thickness. A detector placed on the surface of the material can pick up the temperature rise and output a voltage, which indicates the power absorbed.

The target is angled so that reflections from the target surface do not travel back to the transducer, but miss it and strike the water surface. They will then bounce off the water surface, back down to the target, and be re-absorbed. If reflected ultrasound reaches the transducer it effectively changes the power being delivered, leading to errors as big as 30% in the measurement.

But where to find a suitable absorbing material? As it happens, the Royal Navy already had one, used as a stealth lining on the surface of submarines. Its naval job is to absorb sonar pulses from other ships and submarines so that reflections of the pulses do not give away their position. To do this job the material must absorb within a very small thickness, which is just what NPL needed.

The absorbing material is a polyurethane-based rubber foam of tiny air bubbles, each about 30 μm in diameter. Filler materials are added to make its density a good match to that of water.

The temperature detector on the surface is a very thin (52 μm) layer of the polymer polyvinylidene difluoride (PVDF). PVDF is pyroelectric – that is,

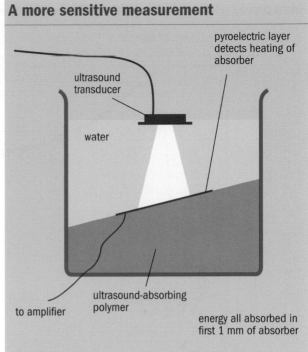

A more sensitive measurement

pyroelectric layer detects heating of absorber

ultrasound transducer

water

ultrasound-absorbing polymer

to amplifier

energy all absorbed in first 1 mm of absorber

The ultrasound beam warms the absorbing material and the pyroelectric layer detects the temperature rise

it generates a small potential difference between its surfaces when the temperature rises. The measurement can be made very quickly because the heated layer of rubber and the detector are both so thin. The ultrasound power only needs to be switched on for a short time.

In fact, the electronics connected to the detector layer are arranged to give an output proportional not to the temperature change but to the rate of increase of temperature. The rate of increase is at its maximum when the transducer is first switched on. Later, energy begins to be dissipated through the water, the membrane or the absorbing material, and the rate of increase slows down. A typical voltage output from the sensor as a function of time is as shown on p228. In this case the transducer was switched on at about 0.8 s. The voltage rises rapidly to a maximum about 250 ms later. The size of this peak is proportional to the incident ultrasound power. The electronics measure and store the magnitude of the peak and convert it into a value of the ultrasound power on the display.

The sensitivity of the method is about 80 mV per watt and is reasonable. When measuring

Measuring ultrasound power by rate of temperature increase of an absorber

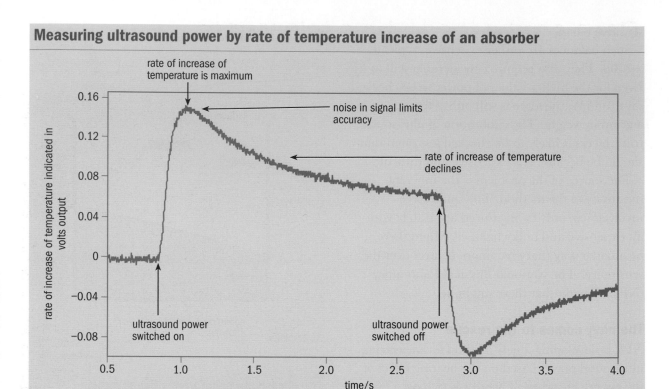

The maximum rate of increase of temperature is proportional to power

This new device makes it easier to measure ultrasound power.

small powers – for example, less than 100 mW – the output may be only a few millivolts, seeming to limit the sensitivity. The answer is straightforward – it is easy to amplify the signal from the detector almost as much as you want, limited only by the presence of noise. In this way the new instrument can measure an ultrasound power as small as about 10 mW, even though this only warms the detector by as little as 10 mK.

You can see that there is random noise in the signal. It is crucial to minimise the level of noise present and there is no profit in increasing the sensitivity by amplifying the signal even more because the noise is amplified as well.

The new device is still under development, but offers several potential advantages: the sensitivity problem is solved; the new device is compact; and measurements can be made quickly and by less skilled operators.

We are grateful to Dr Bajram Zequiri of NPL for assistance in preparing this case study.

Summary

- The sensitivity of an instrument is the ratio of its output to its input
- Improving the sensitivity may require changing the method of measurement
- Outputs in the form of potential differences can be amplified, giving control over sensitivity. The presence of noise limits the useful amplication.
- Other features of an instrument such as response time, ease of use, size and cost are also important

Case study: the Hubble telescope: the most expensive zero error ever

The curved mirror of the Hubble Space Telescope was planned to be more precisely shaped than any previous telescope mirror. And so it was, but the shape chosen was slightly wrong. This was because the instrument used to monitor the mirror shape was set incorrectly, with a lens slightly out of place. Corrective optics had to be fitted to the orbiting telescope before it could start to deliver its stunning results.

Key ideas: systematic error, zero error

This is the story of how a zero error of 1.3 mm in a length measurement cost hundreds of millions of dollars.

The Hubble Space Telescope was launched, after many delays, in April 1990. Astronomers expected to learn much more about objects in the universe from its fantastically clear and sharp images taken outside the disturbing influence of the Earth's atmosphere. Instead, the images they saw were fuzzy, little better than those from a decent Earth-bound telescope. What had gone wrong and could anything be done about it?

The Hubble Space Telescope.

Shaping a telescope mirror

The main mirror of the Hubble telescope was designed to have a diameter of 2.4 m, ground accurately to a curved (hyperbolic) shape. It was designed to be able to be used with short-wavelength ultraviolet light, so the shape had to be exact. The specification called for the mirror to fit

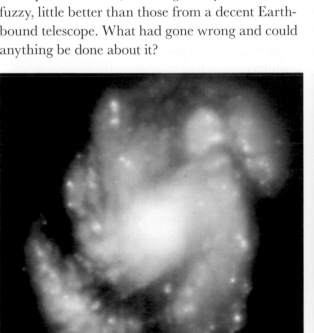

A blurred image from the Hubble taken before correction.

This is an image taken after correction of the mirror.

How the zero error was introduced

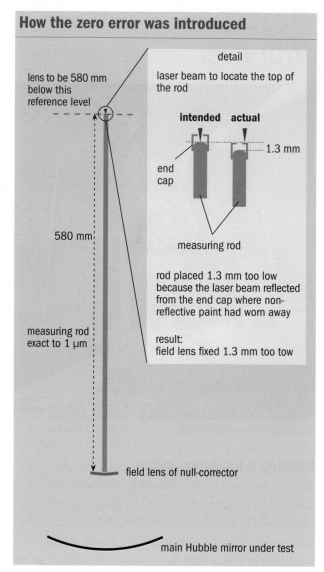

lens to be 580 mm below this reference level

580 mm

measuring rod exact to 1 μm

field lens of null-corrector

detail

laser beam to locate the top of the rod

intended   actual

1.3 mm

end cap

measuring rod

rod placed 1.3 mm too low because the laser beam reflected from the end cap where non-reflective paint had worn away

result:
field lens fixed 1.3 mm too low

main Hubble mirror under test

How do telescope mirror-makers ensure that their mirrors have the right shape? They go through repeated cycles of polishing and testing, gradually improving the shape until the tests show it to be right. These tests are done with what is called a null-corrector, which uses interference of light to judge the correctness of the surface shape. One reason that PerkinElmer won the contract to make the mirror was that it had designed a new kind of null-corrector that promised to be better than existing types. Using the new null-corrector, the mirror was polished again and again until it seemed to be exactly right.

The trouble was that the new, "better" instrument had been incorrectly assembled. To work properly, a lens had to be placed 580 mm below a reference level fixed by the other components of the null-corrector, to a precision of 10 μm. In fact, the lens ended up 1.3 mm below its correct position. How could such a gross error occur?

The answer turns out to be quite simple. To get the spacing right a measuring rod was made, touching the lens at its lower end, with its top supposed to be at the reference level. But the top of the rod was wrongly placed. Laser light supposed to reflect from the top of the rod actually reflected from an end-cap on the rod. So it was the end-cap that ended up in the right place, with the top of the measuring rod 1.3 mm too low.

In consequence, the lens at the bottom of the measuring rod was placed 1.3 mm too low. The assemblers even had to insert extra spacers to position it low enough. Maybe this should have warned them that something was wrong.

There was a good reason to have the end-cap that caused all the trouble. It was important that the laser light hit the exact middle of the measuring rod, so the cap had a small hole centred over the rod to ensure that this occurred. Black anti-reflective paint on the end-cap stopped reflections from its surface, so all should have been well. Sadly nobody noticed that a small chip of paint had worn off the end cap, letting the laser light reflect from the wrong place.

The error built into the instrument used to check the shape of the Hubble mirror meant that it was made perfectly, within the design tolerance of 30 nm, but to the wrong shape.

a specified shape to within $\frac{1}{20}$ of the wavelength of visible light. That is an astonishingly tight tolerance of only about 30 nm, to be achieved over the mirror's entire 2.4 m diameter surface.

The experienced PerkinElmer company, which made the mirror, achieved this remarkable precision using a new system of computer-controlled polishing. Unfortunately the company very carefully gave the mirror slightly the wrong shape. It was flatter at the edges than it should have been. The error was only about 1 μm, which sounds pretty small, but it is more than 30 times the allowed tolerance of 30 nm – not a small error at all. The result was that light from the edges of the mirror did not focus in the same place as light from nearer the centre, making all the images appear fuzzy.

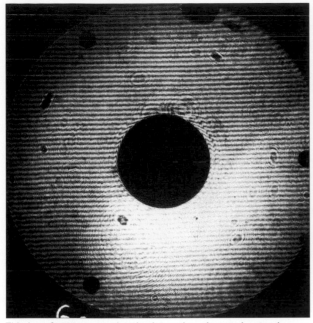

This interferogram spuriously shows the mirror to be good.

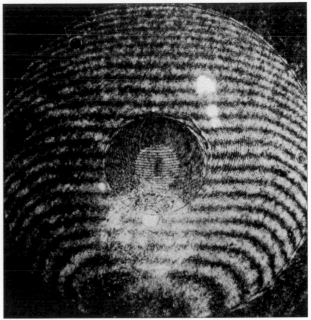

This interferogram correctly shows that the mirror was wrong.

Testing the mirror

The photograph (above, left) shows interference fringes from the faulty null-corrector when used to test the Hubble mirror. The straight, evenly spaced fringes indicate a perfectly shaped mirror but the null-corrector was looking for the wrong shape.

PerkinElmer had two other null-correctors, of a different design, which both showed that the shape was slightly wrong. The photograph on the right shows interference fringes from the Hubble mirror taken at the same time with one of the different null-correctors. The curved fringes show that the mirror shape is wrong. A third check also suggested that the mirror was wrong.

What should one do when measurements disagree? PerkinElmer had to choose which test instrument to believe. The contract guidelines said that the null-corrector used to shape the mirror had to be checked against others but it was less clear about what to do if they disagreed. Take a majority vote maybe? If it had, PerkinElmer would have stayed out of trouble. But why reject the results of its best, newest and most expensive apparatus? The company decided to trust it, as you or I might well have done too. Remember that rejecting the results of the null-corrector used to shape the mirror would have meant starting polishing all over again – months of extra work – and deadlines were tight.

Ironically the mirror then lay in storage for nine years, from 1981 when it was finished to 1990 when the launch took place. The delay was caused by problems with the Space Shuttle, including the tragic *Challenger* disaster. Checks on the mirror could have been made but weren't; remember, though, that these would have been costly.

Giving Hubble corrective spectacles

Despite the mistake the mirror had been given a very exact shape, several times better than the required tolerance. That is, the shape was wrong but to a precisely known extent. This was found out by measuring the defects in the images, by looking at the shaping data kept by PerkinElmer and by testing a back-up mirror made by Kodak.

Thus, entirely because of the great precision of the initial shaping of the mirror, it was actually possible to design further mirrors whose shape would correct the fault. This was done and, at the end of 1993, a 10-day servicing mission took the corrective optics package up to the Hubble telescope and installed it. The opportunity was also taken to replace solar arrays that power the telescope, install new gyroscopes in the system for pointing the telescope, upgrade its computers and electrical systems, and boost the telescope's orbit (which had decayed because of a slight drag from the upper atmosphere).

An image from the Hubble telescope of dust near a star.

An image from the telescope of the Hubble deep field.

The Hubble telescope now began to work as it had always been designed to. It has given astronomers very-high-quality information about all sorts of objects, from nearby star formation to the most distant reaches of the universe. But this would have been achieved three years earlier and for several hundred million dollars less if it hadn't been for a 1.3 mm zero error.

Zero errors are more common than you think

Zero errors occur more frequently than you might expect. Here are some examples.

In precision engineering, accurate spacings are achieved using very accurately machined blocks made in a range of thicknesses. To calibrate a spacing, several blocks may have to be stacked together. Tiny air gaps between the blocks then lead to an error in the total spacing. To avoid this, the blocks have to be wrung together very firmly, to achieve as near perfect a contact as possible.

In your laboratory, you may well have an ultrasound sensor designed to measure distances – an ultrasound tape measure. But just where on the ultrasound device should you take the distance as starting from? Can you see the surface of the ultrasound transducer itself? Perhaps it is covered by a protective mesh. And does the reflecting object have a flat surface at right angles to the ultrasound beam? If not, you may not know just where the ultrasound is reflected from.

Your laboratory may also have sets of micrometer screw gauges or vernier callipers. It is a good exercise to look carefully at their readings when closed up to record zero distance. You may find that the scales do not read exactly zero.

You probably quite often use digital meters to read current or potential difference. Try checking that the meter reads zero when the input is zero, Similarly, what about measuring resistance directly with a digital multimeter? Does the meter read zero when you short-circuit its terminals?

Summary

- A zero error is a fixed error added to or subtracted from the true value. For example, a voltmeter with a zero error will read some small value when it has zero input.
- No amount of averaging of results will remove a zero error. The only thing to do is to correct for it or remove its cause.
- Always check for zero error before taking measurements. For example, look at where the scale begins on your meter rule.
- Don't trust an instrument because it looks new or expensive. Don't trust a sensor just because it is connected to a computer. Always check everything.

Case study: discovering a natural nuclear reactor

To widespread disbelief, French physicists in the 1970s suggested that two billion years ago there had been a natural nuclear fission reactor in uranium deposits in west Africa. A routine measurement of the percentages of different isotopes in uranium ores, with one unexpectedly low value, was the seed for this remarkable proposal.

Key ideas: distribution of values, range, outlying values

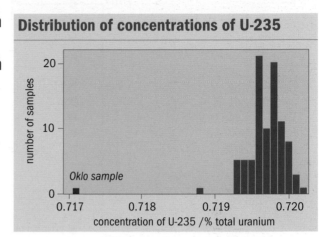

On 2 December 1942 a team led by the Italian physicist Enrico Fermi made the world's first self-sustaining nuclear reactor. Or rather, they thought it was the first. Building the reactor was a huge technological achievement and paved the way for the development of fission reactors in several wealthy and technologically advanced countries. Even now, in the 21st century, building a nuclear reactor is a considerable challenge and one that relatively few countries attempt.

Why then, in 1972, did the French physicist Francis Perrin suggest that there may have been a "natural fission reactor" in the Oklo deposit of uranium ore in Gabon, west Africa? What made him think that nature had achieved the conditions for a self-sustaining reaction in this deposit? After all, Fermi and his team had needed specially enriched uranium, control rods, a moderator and much more besides. What was Perrin's evidence for such an incredible claim?

At first sight Perrin's evidence seems very slight indeed. He was conducting a routine analysis of samples of uranium ores as part of the tracking of all fissionable material imported into France for use in its nuclear reactors. Perrin noticed that a sample from the Oklo ore deposit seemed a bit low in the isotope uranium-235 (U-235). Its concentration was 0.717%, not the usual 0.720% common to most ores. This difference, only 0.003%, does not initially seem very big, but this impression changes when you plot the data

and look at it more closely (see "Distribution of concentrations of U-235" above).

The samples were each measured to ±0.00005% U-235, so the discrepancy was not a problem of uncertainty of measurement. The samples varied over a range so Perrin had to ask himself whether the difference in concentration of the Oklo sample was large or small compared to this variation. The chart shows how Perrin's samples varied.

Have a look at the variation

The chart illustrates a general rule: always plot values to have a look at the variation. Use a histogram (as here) for many values, or a simple dot-plot for fewer (p234). The Oklo concentration lies well away from all the other values. There is a clear gap between the Oklo value and the others, bigger than the range over which the rest of the samples vary.

You might be content with this visual exploration – simply looking to see if there is a big gap – but not all examples are so clear and it is often worth being a bit more quantitative. Just how big is the gap compared with the variation in the values of ore concentrations?

Mean, spread and outliers

An outlier is a value that seems very unlikely to have happened purely as a result of the random influences that make the other results vary.

First plot out the values. To decide whether

Key summary: mean, spread and outliers

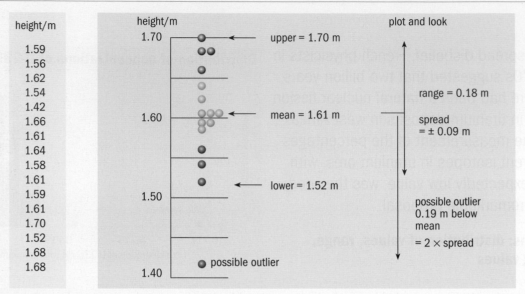

height/m
1.59
1.56
1.62
1.54
1.42
1.66
1.61
1.64
1.58
1.61
1.59
1.61
1.70
1.52
1.68
1.68

what to do:

1 make a dot-plot (or a histogram)

2 look at the distribution of values: identify a possible outlier

3 find the mean, excluding any possible outlier

4 estimate the range of values, excluding any possible outlier

5 take the spread to be ± half the range

6 compare the distance of the outlier from mean with spread

7 decide whether to take the possible outlier seriously

8 IF NOT: recalculate mean and uncertainty limits including "outlier"

Including the "outlier" the mean height = 1.60 m; spread = ± 0.14 m = ± 0.1 m approximately

When results vary, always plot and look. See if possible outliers are more than 2× the spread from mean.

to treat a value as an outlier you need to look carefully at the plot and think about:
- what is the mean value?
- how big is the spread around this mean value?
- is there "clear blue water" between the main cluster of values and the possibly unusual value?

The plot in "Key summary: mean, spread and outliers" shows measurements of the heights of 16 adult married women, taken from a large national survey carried out in the UK, all measured carefully to the nearest centimetre. One value, 1.42 m, does look a little low. Making a dot-plot of the heights lets you have a more careful look. There does seem to be a bit of a gap: there are no cases between 1.52 m and 1.42 m.

The first thing to do is to check that the height was not written down wrongly. This is often the reason why a value looks unusual – it is just

a simple mistake. It would also be sensible to measure again. Let's suppose that there is no problem of this kind in our data. But remember it is always important to check first, before getting into any calculations.

So how big is the gap? The range of the distribution of heights, leaving out the possible outlier for the moment, is from 1.70 m to 1.52 m. So the width of the distribution is about 0.18 m, or a spread of plus or minus 0.09 m. Check using common sense – hold your hands 0.18 m (18 cm) apart. Is this a reasonable range of heights to expect? You will probably decide that it is.

The mean height of these women is 1.61 m, again leaving out the possible outlier for the moment. How much shorter than the mean height is the perhaps unusually short woman? Answer: 0.19 m. This is about twice the spread.

What to decide? We have chosen this example because the answer is not clear. A difference from the mean of twice the spread of the other results is quite rare. That is, this person does seem to be unusually short. However, let's try common sense again. Have you seen any women of about this height? The answer is probably "yes but not often".

If the difference had been three or more times the spread of the other heights, the possible outlier would have had to be taken very seriously. This one is probably just telling us that women vary more in height than you would think from a small sample. So the best solution is to include this unusual value and calculate the mean and spread again. As it happens, a look at the full data for 200 women shows that this is a good decision.

It should be clear to you that these estimates are very uncertain. The spread of a distribution is at best only roughly defined, and will vary depending on which cases are taken. It is never wise to treat an estimate of spread as known to better than one significant figure. If in doubt, round up, not down.

The Oklo example seen up close

Now apply these ideas to the Oklo example. At first sight it looks difficult to estimate the spread because of the second isolated value at 0.7188%. So try it both ways. With this value included, the width of the distribution is 0.0014%. Leaving it out, the width is 0.0010%.

Common sense again comes to the rescue. We are trying to estimate a rough and ready quantity that simply cannot be known accurately. We cannot expect to estimate the spread to more than one significant figure. The spread seems to be between ±0.0005% and ±0.0007%. A reasonable estimate might be ±0.0006%, recognising that this is not to be relied on to better than plus or minus one in the last digit.

The concentration of the Oklo ore deposit is 0.0026% less than the mean, which is more than four times larger than the estimated spread about the mean, ±0.0006%. These estimates confirm the initial visual impression that the gap is really big and that the Oklo data really is an outlier. Calculations something like this were Perrin's first reason to suspect that the Oklo sample was unusual and had to be looked into further.

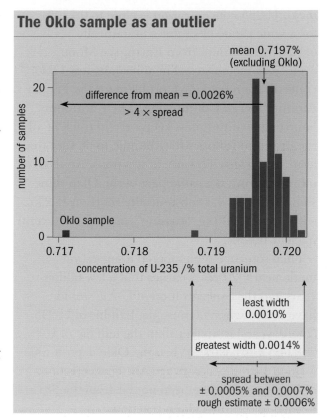

The Oklo sample as an outlier

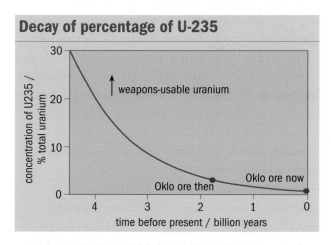

Decay of percentage of U-235

Finding a reason for an outlying value

The analysis says that the Oklo sample certainly ought to be treated as an outlier. But what to do next? Throw that value out as "wrong"? Try to find out if there is a reason for it? The analysis can't tell you what is going on, only that something seems to need explaining.

The first thing to do is to check the measurement itself, but in the case of the Oklo ore deposits this was not the answer. Repeated measurements of the U-235 concentration gave the same low result.

So a physical explanation was needed. The

mean U-235 concentration is close to 0.720% for a wide range of samples of naturally occurring uranium from Earth, the Moon, even from meteorites. The unusually low Oklo concentration became a problem for the French Atomic Energy Commission. Their uranium handling facilities must keep careful track of all the uranium isotopes they handle to be sure that none are diverted for use in weapons. Analysis of more ore from the same part of the Oklo mine showed that it was substantially short on U-235, with around 200 kg "missing", enough for several nuclear bombs.

Then someone remembered a suggestion made nearly 20 years earlier that a few billion years ago the concentration of U-235 would have been higher, because the half-life of U-235 (7×10^8 years) is smaller than the half life of U-238 (4.5×10^9 years). When the Oklo deposit was formed 1.8 billion years ago, the concentration of U-235 would have been not far from the 3% needed to sustain a chain reaction.

The idea was that the Oklo deposit had been the site of a natural nuclear reactor using up some of the U-235 present. But an outlying value and an idea for how to explain it are not enough. You need to test if the idea really works. In this case it meant looking in the Oklo ore for the probable products of a fission reactor. One of these is plutonium, whose fission products were found, including substantial excesses of the isotopes neodymium-143 and ruthenium-99.

Today's picture is that about 2 billion years ago there were 16 natural nuclear reactors in the Oklo region, releasing 15 000 MW-years of energy over hundreds of thousands of years. The average power was, however, quite small, up to 100 kW, which would have raised the temperature of the ore by a few hundred degrees Celsius. Groundwater is thought to have acted as a neutron moderator, leading to a stable chain reaction.

Note of caution

The analysis suggested here is very simple. If you study measurement further you will learn much better and more sophisticated methods of analysing variability in data. These can estimate the actual probability that a value lies outside the main cluster of values. Here all you can do is guess or use a (sometimes fallible) rule of thumb.

Whichever method is used, they all have several things in common:
- variation is thought to arise from many small influences acting randomly and independently – that's why values tend to cluster together;
- a measure of the spread of the distribution is required and several exist. All of them can usually only be estimated very approximately – better than ±25% is rare. The measure of spread used in our example (± half the width of the distribution) is particularly unreliable, and this unreliability can increase if the sample increases. Its main merit is simplicity;
- the presence or not of an outlier is judged by comparing how far it is from the mean with the size of the spread. The question is always: "Could this much difference be the result of random influences acting independently?";
- statistical processing of data is never enough. You always have to add in estimates of uncertainty or variability from knowledge of the instruments used and the things measured;
- refer every decision back to common sense. Ask: "Can it be like this, from what I know?"

Summary

- Always "plot and look": make a visual plot of repeated results or samples. Use a histogram (for many values) or a dot-plot (for fewer values).
- Look for possible outlying values. Double check for mistakes. Use common sense to consider how much variation might be possible.
- Temporarily remove a suspected outlier from the results
- Find the mean
- Estimate the spread from half the width of the distribution
- Don't trust the spread estimate to better than one significant figure and round up rather than down
- Use the rule of thumb that departures from the mean that are greater than twice the spread are to be taken seriously
- Always try to explain any outliers

Questions

1. **This question is about "Case study: the ocean from space" (assume that $c = 3.0 \times 10^8 \, \text{m s}^{-1}$).**
   (a) Calculate the difference in trip time for a radar pulse when the sea level changes by 4 cm. Can this be resolved by the timing system on board the satellite?
   (b) Water vapour in the atmosphere can give an error of 0.5 m in the reading of sea surface height. Calculate the change in trip time this corresponds to.
   (c) Explain why it is important that high-frequency radio waves are used in the altimeter.
   (d) Give an example of a different measurement that may include a systematic error.

2. **This question is about "Case study: calibrating ultrasound transducers".**

   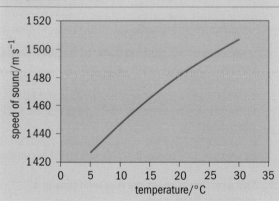

   How the speed of sound in water varies with temperature

   (a) Calculate the force exerted by ultrasound from a 1 W source when absorbed by an object in water at a temperature of 20 °C. The speed of sound in water at 20 °C is 1480 m s^{-1}.
   (b) Use the graph above to estimate the change in the speed of sound per degree rise in temperature over the range 15–25 °C.
   (c) Estimate the uncertainty of the value for the force calculated in part (a) if the temperature of the water is measured to the nearest degree Celsius.

3. **This question is about systematic error.**

p.d./V	I/mA
4.2	19.1
5.1	23.4
6.0	28.0
7.2	34.1
8.4	40.0
9.8	47.1
11.2	54.0

   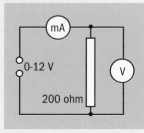

   (a) Explain why a systematic error cannot be corrected for by taking the mean of many readings.

   (b) The table shows results taken by students using the circuit shown. The results do not show the pattern expected and the students disagree about the reasons for this. One suggests that this is because the resistor was not 200 Ω as stated. One suggests that there was a systematic error in reading one of the instruments, while a third student suggests that the internal resistance of the power supply was the cause of the surprising results. Draw a graph of V against I to decide which suggestion is correct. Explain your reasoning.

4. **This question is about the sensitivity of thermocouples and thermistors as temperature sensors. Thermocouples generate an electromotive force (e.m.f.) that can be measured with a suitable meter. The thermistor was used with a 5 V supply and a 1 kΩ resistor to make a potential divider. The table below gives the results of some measurements made with these two sensors.**

Temperature /°C	Thermocouple e.m.f. /mV	Voltage across 1 K resistor /V
0	0.00	1.18
20	0.79	2.22
40	1.61	3.27
60	2.47	4.00
80	3.36	4.42
100	4.30	4.67

   (a) Draw a circuit diagram showing carefully how the potential divider was set up.
   (b) Draw graphs showing how the output of each sensor changes with temperature.
   (c) What is the sensitivity of each sensor at 50 °C?
   (d) Describe what happens to the sensitivity of each sensor as the temperature rises from 0 °C to 100 °C.

5. **This question is about zero errors. The speed of light in an optical fibre can be measured by sending pulses from one end of the fibre to the other and electronically monitoring the pulses at each end. The time delay is usually measured on an oscilloscope but it will be greater than the time taken for the pulse to travel along the fibre. This extra time is an error introduced because the signal takes time to be processed by the electronics. To eliminate this error, two measurements are taken using different lengths of fibre. The time taken for the electronics to respond will be the same for both. In one such experiment, the delay for a 5 m fibre was 0.42 μs and for a 30 m fibre was 0.55 μs.**
   (a) How long did the pulse take to travel 25 m?
   (b) What is the speed of light in the fibre?
   (c) What is the refractive index of the material from which the fibre is made?

6. **This question is about achieving very high precision. Astronomical measurements (astronometry) of the angular position of objects in the sky are made with remarkably small uncertainty. The *Hipparcos* satellite, launched in 1989, measured the positions of several hundred stars to better than 1 milli-arcsecond. In an article, this angle is claimed to be equal to that subtended by the height of a person on the Moon as seen from the Earth. (Distance from the Earth to the Moon = 384 400 km. One degree = 60 arcminutes, 1 arcminute = 60 arcseconds, 2π radians = 360 degrees.)**

(a) What sized object at the distance of the Moon does subtend an angle of 1 milli-arcsecond?

(b) A new satellite, *Gaia*, is proposed. *Gaia* will make measurements to within ± 20 micro-arcseconds. What size of object on the Moon when viewed from Earth represents this improved precision?

7. **This question is about spread and outlying values. Sixteen members of an A level class each had their reaction time measured. One student pressed a hidden switch at a randomly chosen moment, lighting a lamp and starting a millisecond timer. The student being tested had to press a second switch as quickly as possible after seeing the lamp come on. The second switch stopped the timer. The table below shows the measured reaction times for the 16 students.**

Time /ms	
308	400
361	274
318	328
258	265
281	209
257	252
218	140
872	179

(a) Display the values on a dot-plot.

(b) Which value looks like an outlier?

(c) Ignoring the possible outlier, calculate the mean, range and spread of the results.

(d) How far from the mean does the possible outlier lie as a multiple of the spread? Is it sensible to treat the value as an outlier and seek a special explanation for it?

(e) Suggest possible reasons for the existence of this outlier.

(f) Is there any reason in these data to suspect the student who got the low result of 140 ms of having "cheated"?

8. **This question is about the range and sensitivity of an instrument. A student makes an electronic balance by attaching one end of a spring to the slider of a linear potentiometer, with the other end held firmly in a clamp. The potentiometer has a resistance of 10 kΩ and is connected to a 5 V supply. It is made of a uniform strip of resistive material 8 cm long. A digital voltmeter measures the output from the potentiometer as shown and reads to the nearest 0.01 V. The spring has a spring constant of $0.5\,\text{N cm}^{-1}$ and is adjusted so that the slider is at end B with no load (output reading 0.00 V).**

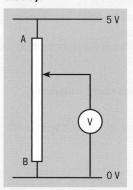

(a) What mass will move the slider to the point A? Assume that the elastic limit is not exceeded.

(b) What will be the output reading with this load?

(c) What mass will give an output reading of 2.50 V?

(d) What is the smallest change in mass that the balance could detect?

(e) It is unlikely that these measurements could be made as precisely in a school laboratory. Suggest reasons why.

9. **This question is about the response time of a measuring instrument.**

(a) Explain why a mercury in glass clinical thermometer has a slow response time of a few minutes. Why can it not be made to respond much faster?

(b) Suggest reasons why the response time of a few seconds for an infrared ear thermometer has advantages for both patient and nurse (see the case study on the CD-ROM).

(c) The response time of a photo resistor to a change in light level is of the order of 100 ms, while that of a phototransistor is of the order of 10 µs. Give practical examples where this difference would, and would not, be important.

(d) Suggest how you might begin to construct a device to record the rapidly changing impulses as raindrops fall on it? Why would it have to be small and light?

Acknowledgments

The publishers are grateful to the following for permission to reproduce images and poetry in this book.

1   The poem "Ultimage Reality" is reproduced by permission of Pollinger Limited and the Estate of Frieda Lawrence Ravagli

1   (right) Courtesy of Dr Alan Cottenden, UCL

5   (left) AIP Emilio Segrè Visual Archives

5   (centre) Courtesy of Mullard Radio Astronomy Observatory

5   (right) [visible light] Digitized Sky Survey, STScI/NASA; [infrared] ESA/ISO/ISOCAM, CEA–Saclay and I F Mirabel *et al*; [radio + infrared] VLA/NRAO (and infrared ISOCAM as stated)

7   (left) John Walsh/Science Photo Library

7   (right) CNES, 1998 Distribution Spot Image/Science Photo Library

8   (left) Courtesy of Matthias Bohringer and colleagues, University of Lausanne

8   (right) AIP Emilio Segrè Visual Archives, Physics Today Collection

9   (top) NCSA/University of Illinois at Urbana-Champaign

9   (bottom) Courtesy of Edward J Groth, University of Princeton

11 Courtesy of AT&T Archives and History Center

13 AIP Emilio Segrè Visual Archives, See Collection

14 (top) Ground-based image courtesy of Georges Meylan (European Southern Observatory); HST image courtesy of Wide Field and Planetary Camera Team and AURA/STScI/NASA

14 (bottom) Courtesy of Professor M Bertero, University of Genoa

17 Lennart Nilsson/Albert Bonniers Forlag

19 (right) Courtesy of Dolland & Aitchison

21 (right) Pavel Hlava, Czechoslovakia

29 (top) Courtesy of Professor Richard S Muller, University of California, Berkeley

29 (middle) David Scharf/Science Photo Library

29 (bottom) Reprinted from J Brugger *Sensors and Actuators A* **34** 193–200 (1995). With permission from Elsevier Science

30 (left) Courtesy of Dr S T Davies, Director of the Centre for Nanotechnology and Microengineering, School of Engineering, University of Warwick

31 Courtesy of Professor William S Bickel, University of Arizona

32 Courtesy of Oak Ridge National Laboratory, Tennessee

34 (top) Dr Jeremy Burgess/Science Photo Library

34 (bottom) Courtesy of Professor Robert Wild, University of California, Riverside

36 STC/A Sternberg/Science Photo Library

43 (top) Donna Coveney/MIT

43 (bottom) Courtesy of Dr A W van Herwaarden and Dr G C M Meijer, Delft University of Technology

45 NASA

46 (top) Courtesy of Professor P N Bartlett, University of Southampton

46 (bottom) AIP Emilio Segrè Visual Archives, Physics Today Collection

47 Courtesy of Professor Julian Gardner, University of Warwick.

48 (bottom left) Previously published in *Microsensors – Principles and Applications*, Julian W Gardner, © John Wiley & Sons Limited. Used with permission

60 Dr Jeremy Burgess/Science Photo Library

75 (top) Science Museum/Science and Society Picture Library

75 (bottom) Courtesy of the Academic Department of Radiology, University of Sheffield

76 (top right and bottom right) The Natural History Museum (London)

76 (bottom left) Supplied by Dulas Ltd (specialists in solar medical equipment)

80 (top) Dave G Houser/Corbis

80 (bottom) Courtesy of Tourism Malaysia

82 (left) © BRE, reproduced with permission

82 (right) Shigio Kogure/Katz Pictures Limited

85 (top) Courtesy of BSI

85 (bottom) Courtesy of Instron ®

86 (three images) Dominic Ross, School of Art and Design, Kingston College

93 (left) Andrew Lambert Photography/Science Photo Library

93 (right) Sheila Terry/Science Photo Library

97 Microfield Scientific Ltd/Science Photo Library

98 (top left) Courtesy of Dr Rodney Cotterill (*The Cambridge Guide to the Material World*)

98 (top right) Andrew Syred/Science Photo Library

98 (middle left) Science Museum/Science & Society Picture Library

98 (middle right) Dr Tony Brain and David Parker/Science Photo Library

98 (bottom left) Courtesy of Dr Claire Davis, University of Birmingham

98 (bottom right) Philippe Plailly/Science Photo Library

101 (top right) Courtesy of SFTC Daresbury Laboratory, reproduced with permission

101 (bottom right) G Müller, Struers GmbII/Science Photo Library

106 Steve Allen/Science Photo Library

107 (left) Bruce Frisch/Science Photo Library

107 (right) Science Museum/Science & Society Picture Library

109 (top) Werner Forman Archive/Wallace Collection, London

109 (bottom) Courtesy of Dr Bernd Kempf, Degussa-Huls AG Dental Division

116 (top) Courtesy of AT&T Archives and History Center

116 (bottom) Astrid and Hans-Frieder Michler/Science Photo Library

123 (top) Robert Picket/Corbis

123 (middle and bottom) From *On the Surface of Things* by Felice Frankel and George M Whitesides, 1997. Published by Chronicle Books, San Francisco. Used with permission

124 (top) Mehau Kulyk/Science Photo Library

124 (middle) Courtesy of Barnabys Picture Library

124 (bottom) Tokyo National Museum

135 David Parker/Science Photo Library

136 (top left) AIP Emilio Segrè Visual Archives

140 (top) Engraving from a painting by Sir Thomas Lawrence, courtesy of AIP Emilio Segrè Visual Archives, E Scott Barr Collection

142 (bottom) Photo Deutsches Museum Munchen

145 (top) University of California

152 (top, set of six images) A Rose *Advances in Biological and Medical Physics* 5 211 (1957)

152 (bottom, two images) Courtesy of Peter Challis, Harvard Smithsonian Centre for Astrophysics

165 (top) Andrew Lambert Photography/Science Photo Library

165 (bottom, two images) C Jonsson *Zeitschrift für Physik* **161** 1961

168 Courtesy of the Archives, California Institute of Technology

171 Courtesy of Dr Robert Reid

173 © Anglia Polytechnic University, Multimedia Motion, Cambridge Science Media

180 Reproduced by permission of Ordnance Survey on behalf of HMSO. © 2008. All rights reserved. Ordnance Survey Licence number 100048011

184 (middle, four images) © Anglia Polytechnic University, Multimedia Motion, Cambridge Science Media

184 (bottom) John Sanford/Science Photo Library

189 (middle) Oxford Scientific Films

189 (bottom) Chris Sattlberger/Science Photo Library

194 (three images) Professor Harold Edgerton/Science Photo Library

199 Ronnen Eshel/Corbis

201 (left) Mike Hewitt/Allsport

201 (right) Malkolm Warrington/Science Photo Library

225 Paul Rapson/Science Photo Library

232 (top) NASA, ESA and H Bond (STScI)

232 NASA, ESA, S Beckwith (STScI) and the HUDF team

Answers

Chapter 1
1. a $+20\,D$, b no numerical answer, c 1.23×10^6 pixels, d $9.0\times6.8\,mm$, e $7.0\times10^{-3}\,mm$ square, f just, if $0.2\,mm$ thick, g $3.7\,Mb$
2. no numerical answer
3. a no numerical answer, b no numerical answer, c $+40\,D$, d $\dfrac{1}{F}=\dfrac{1}{F_1}+\dfrac{1}{F_2}$.
4. no numerical answer
5. no numerical answer
6. a $0.1\,nm$, b no numerical answer, c no numerical answer

Chapter 2
1. a $0.5\,mAV^{-1}$, b $2\,k\Omega$, c $2\,mAV^{-1}$, d $4\,k\Omega$, e e.g., $250\times40\,mm$, f e.g., $125\times10\,mm$
2. no numerical answer
3. a $1.25\times10^{16}\,s^{-1}$, b $1.6\times10^{-15}\,J$, c $20\,W$, d 100 electrons
4. a, b no numerical answer, c (i) $500\,\Omega$, $300\,K$, (ii) $320\,K$, (iii) $0.15\,V$, d no numerical answer
5. no numerical answer

Chapter 3
1. a no numerical answer, b minimum sampling frequency = $2\times$ highest frequency, c about $700\,kbit\,s^{-1}$, d $1.4\,MHz$, e no numerical answer, f $9\times10^{-4}\,mm$
2. no numerical answer
3. a $0.8\,Mbit\,s^{-1}$, b yes, c 1 million pixels, each $0.1\times0.1\,mm$
4. a $0.333–0.375\,m$, b $8000\,s^{-1}$, c 256 levels, d $64\,000\,bit\,s^{-1}$, e $64\,kHz$, f 1500 lines
5. no numerical answer
6. a $2.0\,mV$, b $20\,kHz$

Chapter 4
1. no numerical answer
2. B>A more force, C>B longer, D>C thinner, E>D lower modulus
3. a no numerical answer, b $4000\,N$
4. a $3\times10^{-6}\,m^2$; $1\,mm$, b $600\,N$, c $1\,mm$, d no numerical answer
5. extension = $7.6\,mm$
6. a Young modulus = $2.3\times10^6\,Pa$; uncertainty not less than 6%, probably nearer 10%
7. percentage uncertainty = $\pm0.5\%$; smallest systematic error about $1\,\Omega$
8. a Nichrome, b $1.1\times10^{-6}\,\Omega\,m$
9. 100 times better

Chapter 5
1. a metals, b semiconductors, c metals, d polymers, e ceramics
2. no numerical answer
3. a about $10\,kN$, b incisors: perhaps $1\,kN$; molars much more, c a few newtons, d $20\,kN$
4. no numerical answer
5. no numerical answer

6. no numerical answer
7. no numerical answer

Chapter 6
1. a no numerical answer, b for first order diffraction at $30°$, separation $\sim1\,\mu m$, c separation gets less
2. a 500 wavelengths, b yes, $6\,mm$, c the same ratio, d reduced to $2\,mm$, e slit separation $15\,m$, f no numerical answer, g $0.2\,m$
3. a 15×10^9 rotations s^{-1}, b $0.02\,m$, c same, d no numerical answer, e $10\,m$
4. no numerical answer
5. no numerical answer
6. no numerical answer

Chapter 7
1. a yes, b yes, c, d no numerical answer
2.

Photons

Frequency of phasor rotation /Hz	Energy /J	Wavelength
10^{15}	6.6×10^{-19}	$(300\,nm)$
(300×10^9)	2.0×10^{-22}	$1\,mm$
2.4×10^{14}	(1.6×10^{-19})	$1.25\,\mu m$

Electrons

Momentum mv /kg m s^{-1}	Speed v /m s^{-1}	wavelength /nm
6.6×10^{-26}	7.2×10^4	(10)
(1.0×10^{-24})	1.1×10^6	0.66
1.8×10^{-24}	(2.0×10^6)	0.36

3. no numerical answer
4. no numerical answer
5. a $3\times10^{11}\,mm\,s^{-1}$, b $1\,ns$, c $1\,ns$, d $5\times10^{-11}\,s$ ($0.05\,ns$), e $2\times10^{11}\,mm\,s^{-1}$, f 1.5

Chapter 8
1. a $30\,m$, b $0\,ms^{-1}$, c $25\,ms^{-1}$, d $12.5\,ms^{-1}$, (i) same, (ii) bigger, (iii) same
2. no numerical answer
3. a no numerical answer, b $800\,m$, c $3.5\,km$, d no numerical answer
4. a $1.3\,ms^{-1}$ at $67°$ to bank, b $8.3\,s$, c upstream at $65°$ to bank, d $9.2\,s$
5. a, b, c no numerical answer, d (i) $38.5\,m$, (ii) about $110\,m$

Chapter 9

1 a 44 m, b 7.1 m

2 a $4.2\,\mathrm{ms^{-2}}$, b $4.9\,\mathrm{ms^{-2}}$, c $8.4\,\mathrm{ms^{-2}}$

3 a $13.4\,\mathrm{ms^{-1}}$, $26.8\,\mathrm{ms^{-1}}$, $31.3\,\mathrm{ms^{-1}}$, b $2.7\,\mathrm{ms^{-2}}$, c 3.1 kN,
d $8.1\,\mathrm{ms^{-2}}$, e 9300 N, f $7.7\,\mathrm{ms^{-2}}$, g 8860 N, h no
numerical answer

4 a, b no numerical answer, c $3.2\,\mathrm{ms^{-2}}$, d 1.6 m

5 a $v_\mathrm{H} = 20\,\mathrm{ms^{-1}}$, $v_\mathrm{V} = 34.6\,\mathrm{ms^{-1}}$, b no numerical answer,
c 140 m

6 a $3\,\mathrm{ms^{-2}}$, b 20 s, c 39 kN, d 23 MJ, e no numerical
answer, f 1.2 MW, g no numerical answer

Case studies

1 a 0.3 ns, yes, b 3.3 ns c, d no numerical answer

2 a force $= 6.76 \times 10^{-4}\,\mathrm{N}$ (equivalent to 69 μg),
b $\Delta v = 1495 - 1465 = 30\,\mathrm{ms^{-1}}$, $\Delta T = 25 - 15 = 10\,^\circ\mathrm{C}$
so change in speed per $^\circ\mathrm{C} = 3\,\mathrm{ms^{-1}\,^\circ C^{-1}}$, c $3\,\mathrm{ms^{-1}}$
in $1480\,\mathrm{ms^{-1}}$ is 0.2% uncertainty

3 a no numerical answer, b graph shows a non-zero
intercept suggesting a zero error of 0.4 V in the
voltmeter reading or a 3 mA zero error in the ammeter
reading

4 a, b no numerical answer, c thermocouple $43\,\mathrm{\mu V\,^\circ C^{-1}}$,
thermistor $37\,\mathrm{mV\,^\circ C^{-1}}$, d no numerical answers

5 a 0.13 μs, b $1.9 \times 10^8\,\mathrm{ms^{-1}}$, c 1.6

6 a 1.8 m across b about 40 mm across

7 a no numerical answer, b 872 ms, c mean 270 ms,
range 260 ms; spread ±130 ms, d 4.6 times the spread;
appropriate to treat as outlying, e, f no numerical
answers,

8 a 400 g, b 5.00 V, c 200 g, d 0.8 g

9 no numerical answers

Index

Page numbers in *italics* refer to entries mainly or wholly in photographs, diagrams or summaries.

plus scale *11, 13*
polarisation 66–7
polycrystalline 100, *101*
polyester resin 105
polymers 77, 113, *114*
    electrically conducting 46
    fibre-reinforced 75, *76*
    properties 86, *118*
    strengths *82*, 83
polythene *113, 114*
position sensor 40
potential difference 31–2, *34*, 35–7
    controlling and measuring 39–42
    in sensors 43–5, 49
potential divider 37, 39–41, 46
    strain gauges in 47, 48
potential energy 31–2, 211–12
pottery 111
power
    calculating 32, 36
    force and velocity *214*
    for a high speed train 213–14
    of a lens 20, 22
    ultrasound transducer 225–7
pregnancy 1–4, *225*
probabilities 152, 154, 156, 166, 168
projectiles 200–2
pulsars *5, 6*
pulse speed *3, 222*
PVDF 227
pyroelectric 227
Pythagoras theorem *178*, 180

Q
quantum behaviour 151–70
    calculations 155–6
    electrons 165–6
    energy 154–5
    of light 152–4, 158–64
quantum physics 145, 147
quartz crystal 43, 44, 45, 49

R
radar 53, 147, 190–1, 192
    altimetry 220, *221*, 222
radiation sensor *43*
radio 39–40, 58, 65–6, 68
    astronomy 5–6, 147
    waves 154–5
railway gradients 211–12
randomness 153–4
rays 19, *20*, 22–3
reflection *127, 138*, 158–9
refraction *138, 160*, 161
refractive index 20–1, *22*
relative directions *176, 177*
relative velocity 191–2
resins 105, 106–7
resistance 35–7, 39–40, 89, 93
    air resistance 199, 200

in sensors 43–5, 47, 49
resistive track 39, 40
resistivity 88–9, 93
resistors 39–40, 41
resolution
    of an image 2, 6, 7
    high resolution 220–1
    of optical instruments 147
    sensor 51
    vectors 177, *179*, 180
response time 45, 51
resultant phasor *155, 156, 158, 160, 161*, 165
results, variations in 41, 91, 233–4, 236
retina *17*, 18, 22, 153
rivers *34*, 35, 36, 37
robots *43*
rods *17*, 18, 153
    aerial rods 66, 131
Römer, Ole 133–5, 136
rough estimate 92, 93, 221, 235
Royal Navy 227–8
rubber 114
rule for lenses 22–3

S
salt *112*
sampling 61–2, 63
satellites 7, 66, 71
    finding the satellite's position 223–4
    monitoring changes to the oceans 220–4
scalars 176, 182, 204
scanning electron microscope *60, 92, 97, 98, 99*
scanning tunnelling microscope *8*, 9, 97, *98*
scattering 66
Seebeck effect 43
seiche 131
semaphore 57
semiconductors 49, 76, 88, 116, *117*, 118
sensing
    controlling and measuring potential differences 39–42
    making microsensors 29–33
    measuring well 51–4
    miniature circuits 29–30, 34–8
    sensors and our senses 43–50
    sensor 43, 51
    ultrasound transducer 226–8
sensitivity 51, 226–7, 228
sensors *see* microsensors
series
    circuits 37, 40, 49
    rivers 36
Shannon, Claude *11*, 69
shaping *110*, 111
ships 192

signalling 57–74
signals
    components 67
    from sensors 43, 47, 49
    spectrum of 68, 69
    time-division multiplexing 67–8
    see also analogue signals; digital signals
silica 77, 103, *112, 123*
silicon *29*, 30–1, 34, 100, 116
    doping 118, *119*
    solar cells 76
silicon nitride 111
silicon p–n junction 45
sintering 111
skiing in Scotland 180
skyscrapers *80*
sliding contact 39–40
slipping *110, 114*
soap film 124, 127, 129
solar
    cells 76
    spectrum *142*
sound 67, 125, 129–31
    digitising 61–3
    frequency *1*, 3, *4*
    waveforms and spectra of *67*
South Pole journey 171
space *see* galaxies; stars; universe
spacings, accurate *93*, 232
spectrum
    electromagnetic 5, *6*
    and gratings 142, *143*
    of the signal 68, 69
    solar *142*
speed 172–4, 176, *182, 183*, 197, *207*
    and braking distances *209*, 210
    see also velocity
speed of light 21, 133–5, 138
    uncertainty and *52*, 53, 224
speed–time graphs 174, *196, 198*
spider's web 81–2
sprint start 204–5
square-wave signals 69
stable interference 129
standing waves *128*, 129, *130*, 131
stars 6, *14, 184*
steel 81, *82*, 85, *86, 98*
    properties 103, *104*, 106
    Young modulus 92
stiffness 83, 84, *113*
stone, falling 199–200
stopping distances 198, *209*
storm cloud 31
strain *83, 84, 85, 86*
    energy 103
    gauge 47, *48*
strengths of materials *82*, 83, 103–8
stress 81–4, *85, 86*
    problems in measuring 91–2

Index

sulphur 114
superconductors 75, 88
superposition 124–9, 140
    quantum superposition 155, 162
systematic errors 51–2, 53, 93
    correction and removal 222, *223*
    easy to make and hard to detect 224
    and high resolution 221
    see also error correction; zero error

T
take-off 204
telephone 58, 61, 62, 67–8
telescopes 5, 6–7, 18, 147, 162, 163
    Hubble Space Telescope *14, 45,*
        229–32
television 58, 66
    digital 69, 71
tennis 200–1, *207*
tensile
    strength *104,* 105, *105,* 106
    testing machine *82,* 83, *85*
tension *77,* 80, *81*
thermal sensing 46–7
thermistors 47
thermocouples 43, 45
thermometers 45
thinking distance 198, *209,* 210
thrust 204
timber 80, 81
time
    and distance 171–3
    distance–time graphs *172, 173, 196,*
        198
    measurements of 220–1
    speed–time graphs 174, *196, 198*
    and velocity 182, *197*
    see also trip times
time-division multiplexing 67–8
times scale *11, 13*
tin dioxide *47*
Tonomura, Akiro *166*
tools 77–8
toughness of materials 78, 103–8
traffic incident 195–6
traffic lights *22*
trains
    going uphill 211–12
    powering a high speed train 213–14
transducers 225–8
transistors 116, 118
transmission grating 142, *163*
transport 172
    engineering 209–14
    see also aircraft; cars; trains
transverse waves 66
tree house 80
trip times *155,* 156, *158, 159, 160, 161,*
    162
tunnelling current *8, 9*

U
ultrasound 1–4, 171–2, 225
    tape measure 4, 232
    transducers 225–8
ultraviolet light 142, 229
uncertainty 41, 51, 91, 92
    speed of light and *52,* 53, 224
uniform acceleration 194, *197*
universe, images 5–7, *9, 14,* 18–19, *184*
uranium 233, 235–6

V
valves 116
variations in results 41, 91, 233–6
vectors 171, 176–81, 204
    adding vectors *178, 183,* 192, 207
    notation *178*
    quantities 176, 184, *185*
    resolving 177, *179,* 180
    velocity 182–4
velocity 182–6
    and acceleration 194, *195, 196, 197*
    changing from forces 206–7
    components of 183–4
    force and power *214*
    relative velocity 191–2
    vertical and horizontal components of
        200, 201, *202*
    see also speed
vernier callipers *93,* 232
vibrations 129, 131
viscose 85–6
visible light *5,* 6
voices 131
voltage variations 70–1
vulcanisation 114

W
water current *34,* 35, 36, 37
water waves 65, 124, 137, *145*
wave behaviour 123–49
wave theory of light 140, *141,* 142, 145,
    168
wavelength
    in the electromagnetic spectrum 5, *6*
    from fringe spacing 140, *141*
    radio 65, 66
    ultrasound pulse 3, *4*
wavelets 137–8, 145, 147
waves 19–20, 21, 22–4
    from apertures 145, *146,* 147
    colours of thin films *127*
    electromagnetic 65–72
    Huygens ideas on 136–8
    phase differences 125–7
    power and force of 226–7
    standing waves *128*
    superposition 124–5
    transverse 66
weight 205

wind instruments *130,* 131
wood 80, 85–6, 97, *98*
work 211
work hardened 111

X
X-ray crystallography 99
X-rays 1, 142, 154, 155

Y
Yew wood 75–6
yield stress 81, *82,* 83
Young, Thomas 140
Young double-slit experiment 140, *141,*
    155–6, 165
Young modulus 83–4, 92, 106, *113,* 140

Z
zero error 51, 52, 229, 230, 232

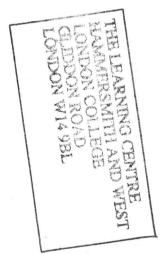